IMAGES
of America

HAMILTON'S INDUSTRIAL HERITAGE

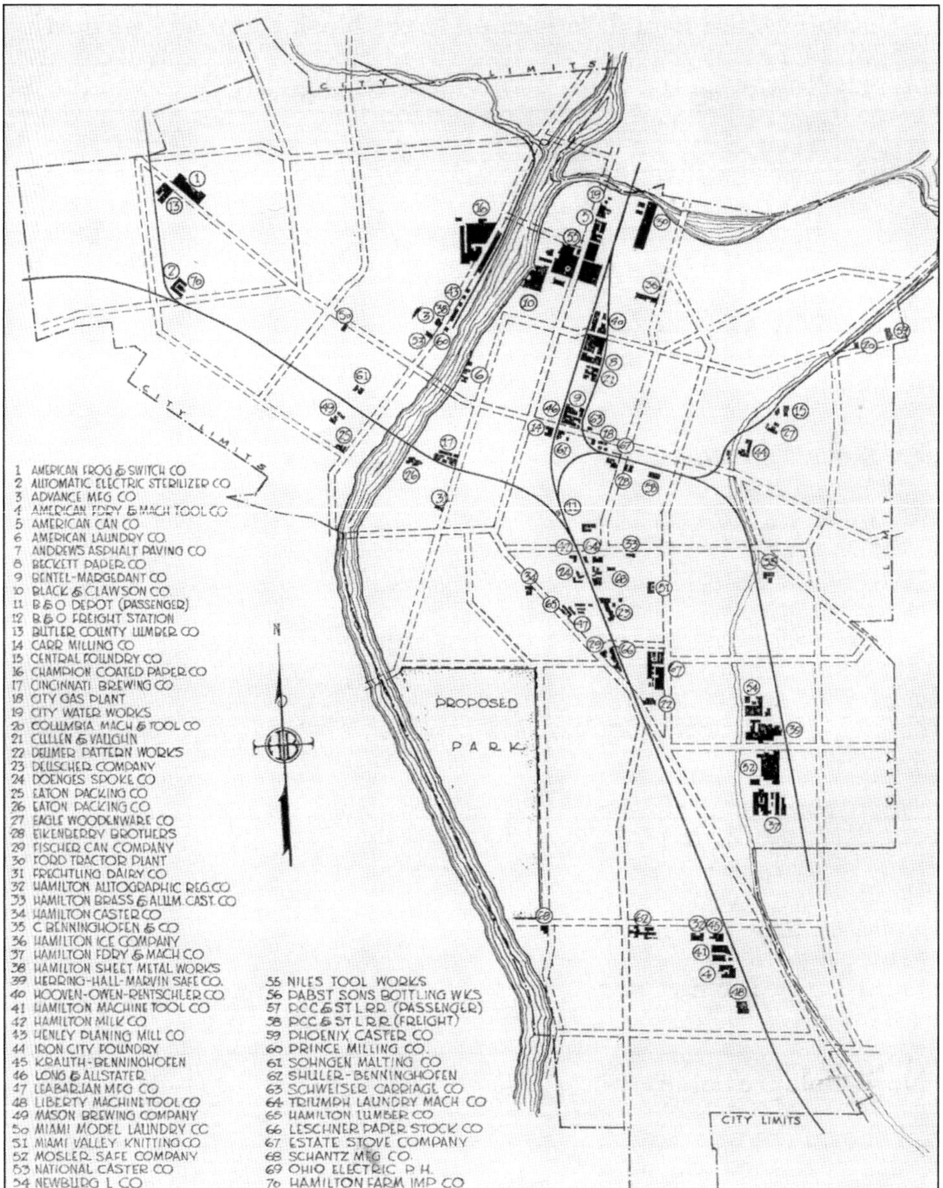

This map from the City Plan of Hamilton, developed in 1920 by city plan engineer Harland Bartholomew, shows the location of the major industries then operating in Hamilton. The plants of more than 100 companies, concentrated along the Great Miami River, the old hydraulic system, and railroads serving the city, made Hamilton a national leader in the manufacturing of Corliss engines, machine tools, paper, safes, vaults, and a wide variety of other products. (Courtesy of the Butler County Historical Society.)

On the Cover: Barrels for moving parts, overhead shafts, and pulleys for power enabled the workers of Hamilton Caster & Manufacturing to produce the firm's single- and double-wheeled casters in the early 1940s. This small family-owned company, formed in 1907 and incorporated in 1910, has survived for more than 100 years and is the oldest manufacturing firm still operating in Hamilton. (Courtesy of Hamilton Caster & Manufacturing Co.)

IMAGES
of America

HAMILTON'S INDUSTRIAL HERITAGE

Richard N. Piland

ARCADIA
PUBLISHING

ISBN 978-1-4671-1379-3

Published by Arcadia Publishing
Charleston, South Carolina

Printed in the United States of America

Library of Congress Control Number: 2014950963

For all general information, please contact Arcadia Publishing:
Telephone 843-853-2070
Fax 843-853-0044
E-mail sales@arcadiapublishing.com
For customer service and orders:
Toll-Free 1-888-313-2665

Visit us on the Internet at www.arcadiapublishing.com

To the visionary men and women who worked to make Hamilton "known in the world's markets" and the "greatest manufacturing city of its size in the world"

CONTENTS

Acknowledgments

I would like to thank the many people who had a hand in making this book possible. Their kindness and support for the project is greatly appreciated. I first must thank Kathy Creighton, executive director of the Butler County Historical Society, for giving me unlimited access to the society's large photographic collection and the wonderful archival resources housed in the Olive S. and Herbert T. Randall Research Center. I also want to thank Kathy and volunteers Richard Scheid and Marlene Carmack for the help they gave in identifying sources of background information and helping with some of the research. They should know that their contribution to the project is truly appreciated.

I want to express my deep gratitude to Susan Myers for providing me with some of the photographs in the George Cummins collection that have not yet been released to the Lane Library. Her thoughtfulness and support for work dealing with Hamilton's local history is very much admired.

Unless otherwise noted, all images used in this book appear courtesy of the Butler County Historical Society. Much of the society's collection was donated by people who worked for the companies mentioned in the book or from the manufacturers themselves. I am grateful they preserved the materials.

Additional contributions to the project were made by several people, and I want to thank the following for the photographs they provided: David Douglas Buckner, Jackie Freyberg, Richard Haid, David Lippert, Steven Lippert, Susan Myers, and Rob Wile. I regret that I could not use all of the photographs I was offered.

Several people gave me great encouragement for my work. I greatly appreciate the friendly support provided by Dave Belew, Joni Copas, John Haid, Richard Haid, and Daniel Lissman. I want to thank Julia Simpson, my editor at Arcadia Publishing, for taking this project under her wing. She gave me great support and inspiration. I hope she thinks the book was worth her efforts.

Finally, I want to acknowledge the support and understanding of my wife, Marcy. She has always been a source of constant encouragement. For 40 years, she has made my life wonderful.

INTRODUCTION

Since its founding in 1791, Hamilton has nearly always been an important center of activity. After Ohio became a state in 1803, the city became an agricultural hub and seat of justice for Butler County. Farming was a major part of the local economy, and farmers were able to ship their crops and livestock via the Miami-Erie Canal, which connected Hamilton to the markets in Cincinnati and northern Ohio beginning in 1827.

The town's location on the Great Miami River attracted a variety of millers, who wanted to power their businesses with the river's current. By the late 1830s, dozens of mills dotted the landscape. Hamilton's industrial history really began in 1845, when the Hamilton and Rossville Hydraulic Company diverted the river's flow through town and developed a system that supplied cheap waterpower to area mills. The hydraulic generated enough horsepower to attract mills that produced paper, flour, cotton, wool, and machine shops. Other early businesses included sawmills that cut lumber for constructing homes and businesses, planning mills that made flooring and weather boards, cotton mills that made yarn and fabrics, grinding mills that broke stone, and foundries that melted iron for castings.

By the 1860s, Hamilton became a leading producer of farm implements such as reapers, harvesters, hay rakes, and plows. The advent of steam engines contributed to a seismic shift in production, and the population of skilled artisans and workingmen in the community brought new factories to the city. By 1900, Hamilton was "the greatest manufacturing city of its size in the world" and produced a staggering variety and diversity of products for the world's markets. The city had at least 28 foundries that made everything from stoves, steam engines, agricultural implements, machine tools, furniture casters, ornamental fencing, and bank vaults.

The men who headed the manufacturing plants also worked to build the community into a viable and vibrant town. Clark Lane of Owens, Lane and Dyer donated the building and 2,000 books for Hamilton's public library. Moses Mosler of the Mosler Safe Company and Lazard Kahn of the F. & L. Kahn & Bros. stove foundry were among the businessmen who worked to develop east Hamilton into a solid residential area and attract new companies to relocate there. Many of the corporate leaders served on the public school board of education, aided in building community-based facilities such as the YMCA, and facilitated modernization of public utilities.

Hamilton's factories advanced, developed, and became the preeminent leaders in their business segments and the standard by which their competitors were judged. In the 1940s, Hamilton was home to several of the world's largest industries. The Champion Paper Company was the world's largest coated paper mill. More than half of the world's safes and vaults were made by the Mosler and Herring-Hall-Marvin safe companies. Niles Tool Works was one of the largest machine tool manufacturers, Hooven-Owens-Rentschler was one of the largest Corliss engine builders, and Estate Stove was one of the largest stove makers in the world.

More than 150 factories and shops produced a staggering variety and diversity of products for the world's markets. Many of the manufacturers were never incorporated or operated on a small scale. Some did not survive beyond one or two years. Others closed their doors or were sold to larger companies. Several moved away to other cities.

It is not the goal of this book to catalog all of the manufacturers that ever existed in Hamilton. Nearly 50 of the companies that had a strong presence in the community are discussed in the following pages. The companies have been grouped into six chapters as a matter of convenience. Chapters are devoted to companies that manufactured paper and the three firms that built bank safes and vaults. A third chapter focuses on seven companies that built engines and vehicles, while a fourth presents another seven firms that manufactured machine tools. A fifth chapter focuses on eight important foundries. The last chapter presents 20 companies that manufactured a wide variety of other products. Several of the featured companies will be familiar to many readers; some will be completely unknown. But they all contributed to making Hamilton a major industrial power that was known for providing excellent products.

Many writers have published helpful accounts of Hamilton's history, some in the context of telling the story of Butler County, and others in recounting the development of the community, an industry, or a company. For this book, the most helpful were the following: two books by Stephen D. Cone, *A Concise History of Hamilton, Ohio* (1901) and *Biographical and Historical Sketches, A Narrative of Hamilton and its Residents From 1792 to 1896* (1896); the three-volume *Memoirs of the Miami Valley* (1919), edited by John C. Hover, Willard J. Wright, Joseph D. Barnes, Clayton A. Leiter, Walter D. Jones, John Ewing, Charlotte Reeve Conover, and W.C. Culkins; and *Centennial History of Butler County* (1905), edited by Bert S. Bartlow, W.H. Todhunter, Stephen D. Cone, Joseph J. Pater, and Frederick Schneider. Special notice should be given to the books published by local historian Jim Blount. Several of his titles, as well as his many columns for the *Hamilton Journal News* and, more recently, the Lane Library, have been especially useful for this book. Specific information concerning the dates when some of the companies included in the book were incorporated, merged, and dissolved was found in the business records kept by the Ohio secretary of state.

One

The Hydraulic and Water Power

In 1840, Henry S. Earhart devised a scheme by which water could be diverted from the Great Miami River and channeled through Hamilton to provide power for mills and shops. Engineer John W. Erwin surveyed and mapped a route and estimated the cost for the work. On March 25, 1841, the Ohio Legislature passed an enabling act for the "Hamilton and Rossville Hydraulic Company" and authorized it to build a dam across the river and a canal to the town to deliver waterpower to mills and factories along its course. The directors of the company, elected in January 1842, included William Bebb (president), Lewis D. Campbell (secretary), Henry S. Earhart (treasurer), Jacob Hittel, Jacob Matthias, Andrew McCleary, Laomi Rigdon, and John Woods. Another hydraulic was built in Rossville in 1850, but it did not have as much success as the Hamilton system.

Construction of the hydraulic was completed on January 27, 1845, when the first water passed through the system. Beginning at the company-built dam four miles north of Hamilton, the diverted water flowed through an excavated canal to a large reservoir that covered about 70 acres, then to a small, seven-acre reservoir on the north edge of town. The hydraulic split into two branches, one going west along Mill Street toward the river and turning south to run parallel with the river. The main canal continued south along North Fifth Street to Stable Street (now Market Street), where it turned to the west and flowed back to the river. The total fall of water from the top of the upper hydraulic to its reentry near High Street was 29 feet.

The hydraulic was a speculative venture that proved to be a resounding success. There were virtually no mills or factories in the areas where the hydraulic canals were built. The lure of relatively cheap power attracted mills to process paper, cotton, flour, and wool, as well as machine shops. The hydraulic remained a primary source of power for local industries until the 1880s, when coal-fired steam engines became economical. Clearly, the hydraulic was the most important factor in the early development of Hamilton as a manufacturing center.

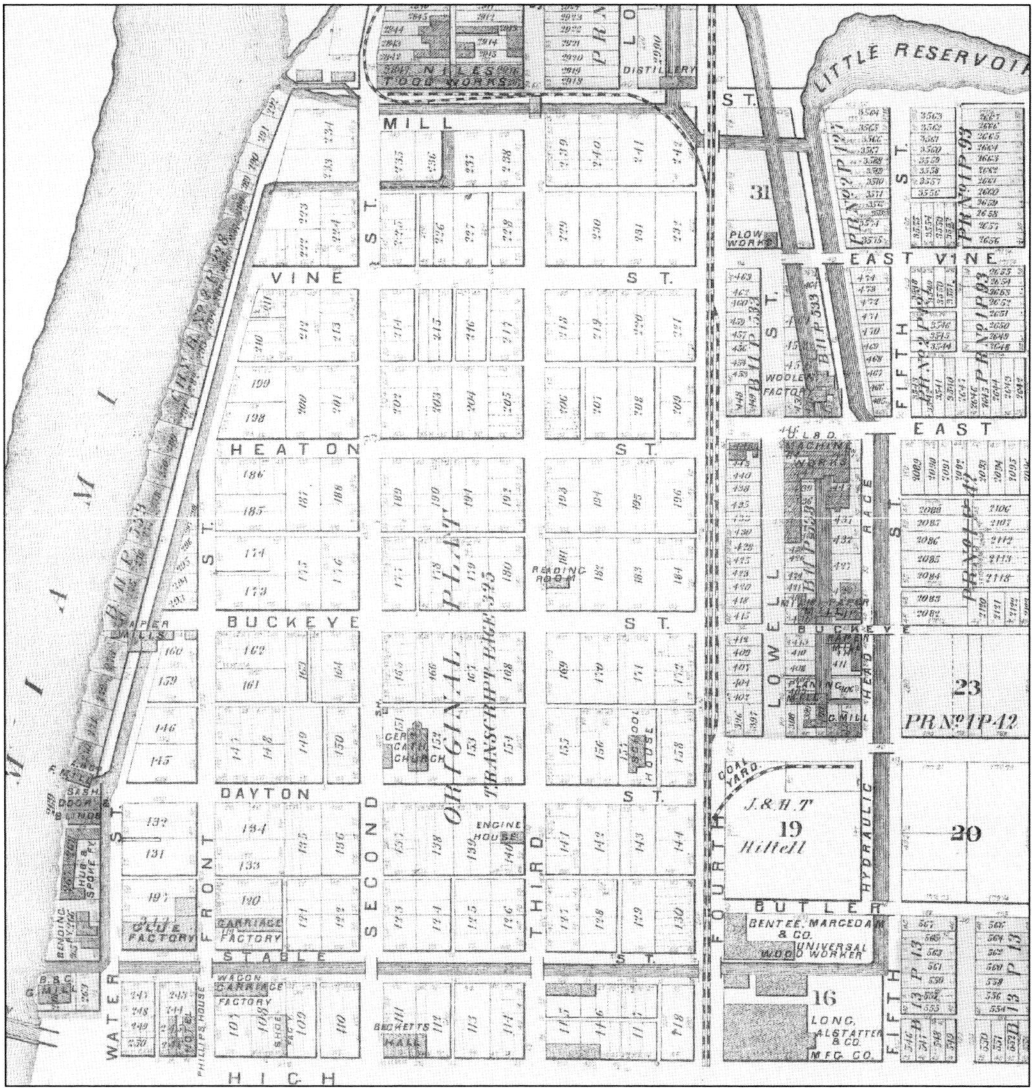

This detail from an 1875 Butler County atlas shows the canals that made up the hydraulic. The water flowed from the little reservoir southward along North Fifth Street until it reached Stable Street, where it turned west and emptied back into the river. Another branch ran along Mill Street before flowing southward to the spillway at Stable Street. It was along these canals that manufacturers built their mills and facilities. Water ran over millwheels that were connected to gears, wheels, and pulleys, providing the power needed by the businesses on the hydraulic. Canals were roughly five feet deep and about forty-five feet wide at the waterline. Water taken from the hydraulic and used by businesses was discharged to a tailrace that carried it back to the river north of the city. Taken together, almost two miles of the total length of the hydraulic system could be used to power industries along its course.

The start of the hydraulic is seen in this photograph. The Little Reservoir is at upper right. The hydraulic canal is flowing under a bridge on Vine Street in the center foreground. A portion of the Sohn and Rentschler foundry can be seen at the upper left, just in front of the hydraulic box. The configuration of the railroad tracks indicates that the photograph was taken sometime after 1895.

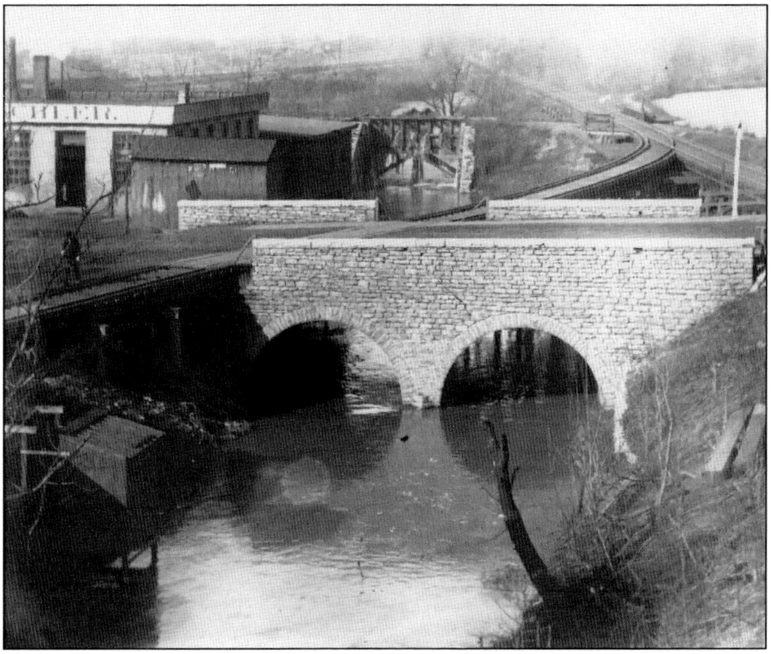

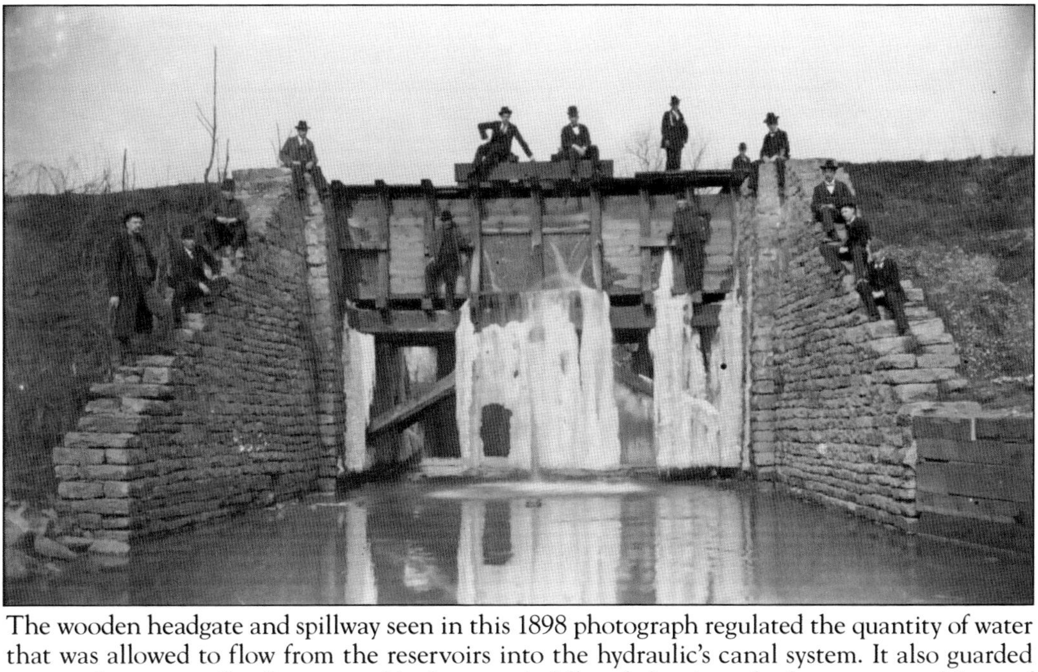

The wooden headgate and spillway seen in this 1898 photograph regulated the quantity of water that was allowed to flow from the reservoirs into the hydraulic's canal system. It also guarded against destructive flooding caused by heavy rainfall and thawing of a frozen stream. None of the men in the photograph are identified.

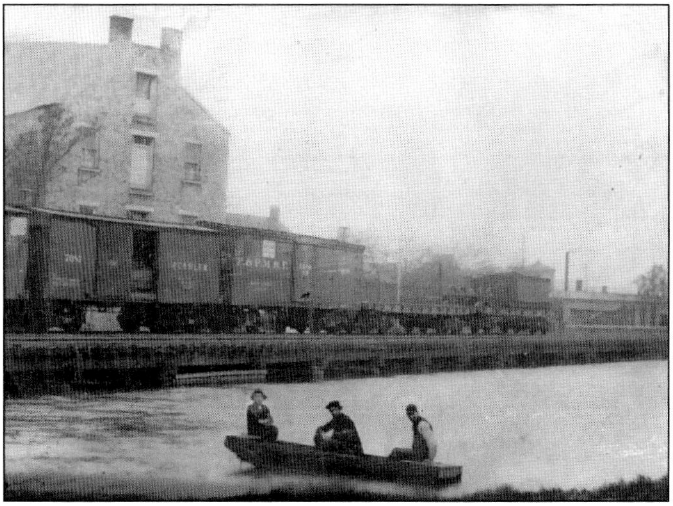

Seen behind the train in this c. 1885 photograph are two early businesses. The Sohn and Rentschler foundry is the low building on the far right, and the Woodrough, Dunn, and McParlin sawmill is the tall building. The sawmill was originally built as a cotton mill in 1845 by William Bebb and Lewis D. Campbell, two of the hydraulic company's officers. The men seated in the boat on the hydraulic are, from left to right, Henry Filmore, an unidentified person, and Alex Cavenaugh.

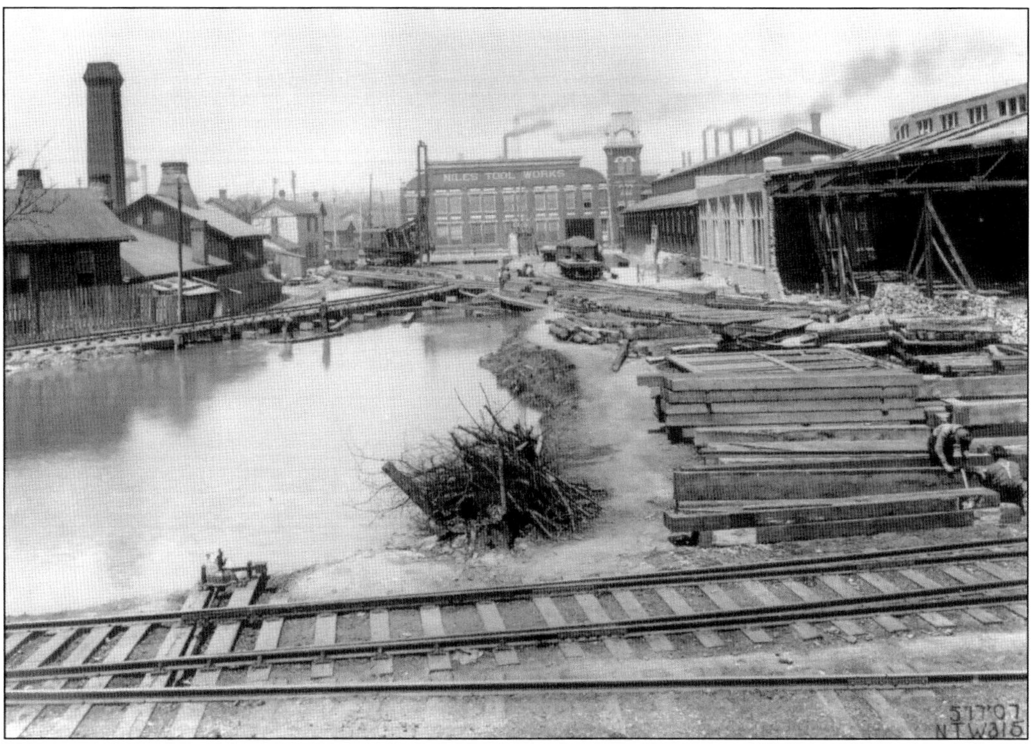

This photograph, taken in May 1907, shows that the Niles Tool Works, located on Third Street, was very close to the branch of the hydraulic that ran along Mill Street. At left are the kilns and smokestack of the Hamilton Tile Works, located one block away at the corner of Mill and Fourth Streets. The railroad tracks are a spur of the Cincinnati, Hamilton & Dayton Railroad, which served both companies and other industries in the area.

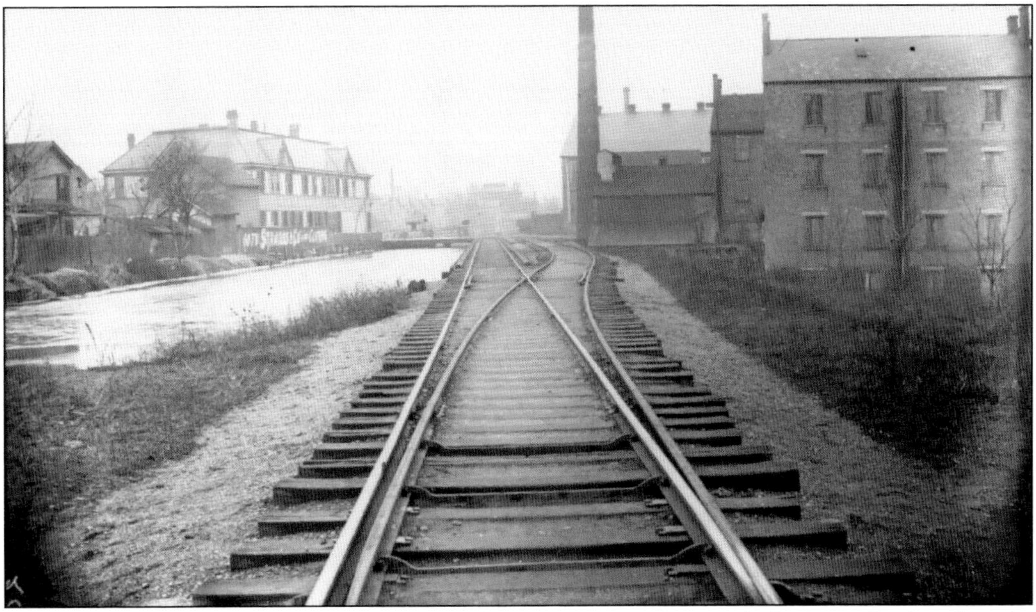

This view of the hydraulic and the Pittsburgh, Cincinnati, Chicago & St. Louis Railroad tracks looks down Fifth Street, north of Vine Street. The Miami Woolen Mills, operated by Shuler and Benninghofen, was in the building in the right foreground. Across the hydraulic was the Lakeview Hotel on the corner of Fifth and Heaton Streets. During the March 1913 Great Miami flood, the hotel collapsed. At least 50 people died when the building's foundation was swept away.

This artist's rendering in an 1875 Butler County atlas depicts the Miami Paper Mill, operated by the Beckett, Laurie and Company. The mill stood on Buckeye Street at the hydraulic. The firm, founded in 1848, was one of the earliest manufacturers to establish a business on the hydraulic. The company's mill remained an important Hamilton employer for more than 163 years, until it closed in 2011.

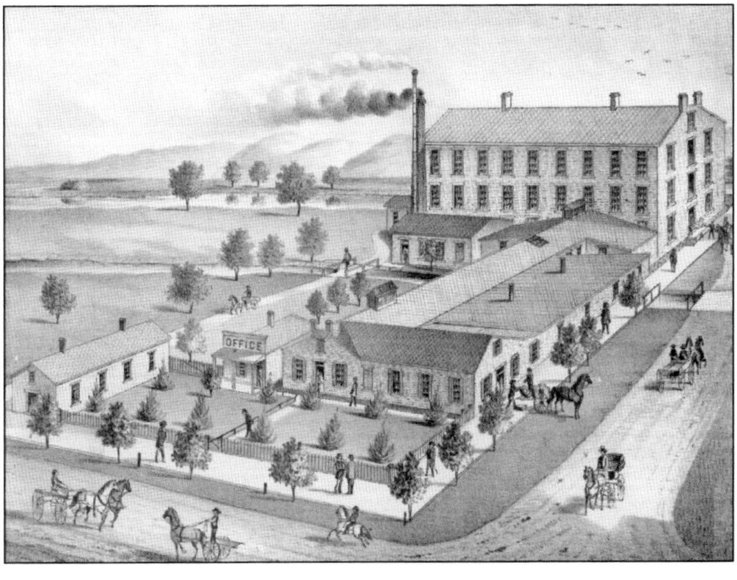

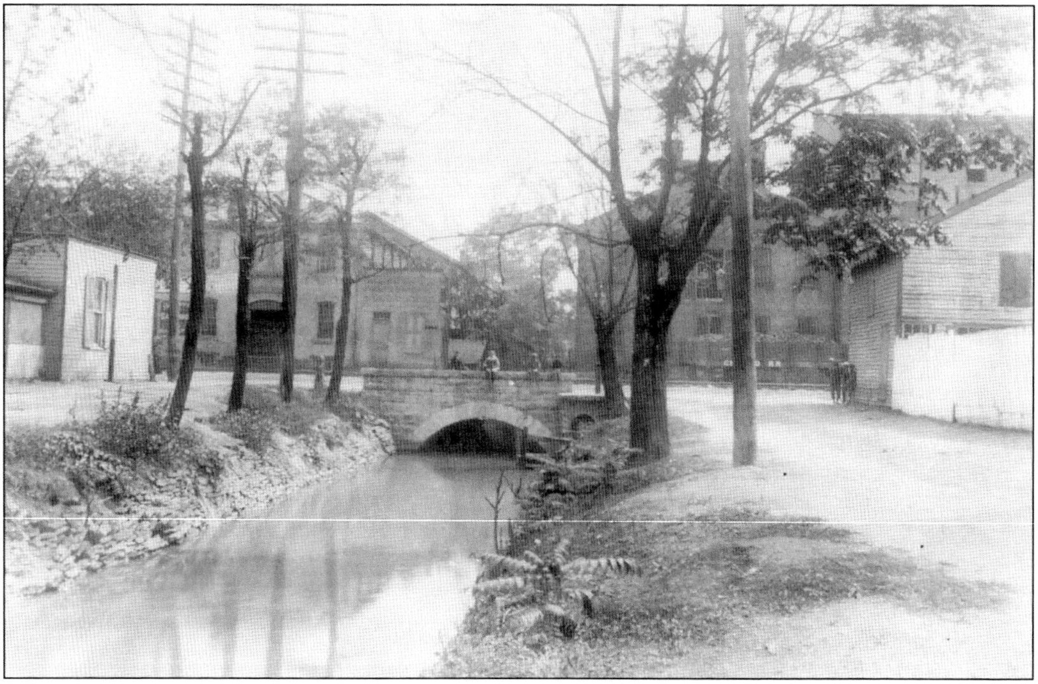

This 1890 photograph of the hydraulic looks east down Market Street toward Fourth Street. Visible beyond the bridge are two important businesses. The building on the right was part of the Long and Allstatter Company's farm-implement plant. On the left was where the Bentel and Margedant Company manufactured its line of woodworking machinery. In the early 1900s, the hydraulic was enclosed in conduits and buried out of public view.

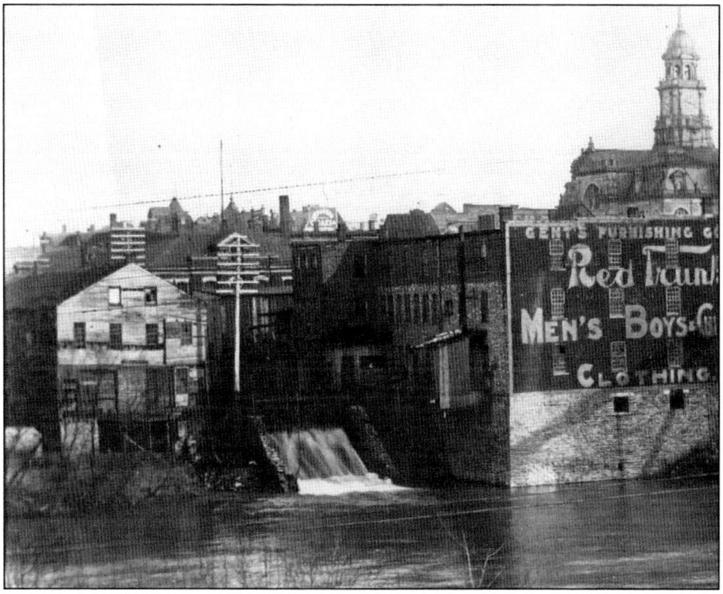

Water in the hydraulic flowed down the canal in the middle of Stable Street until it reached the western end and emptied into the Great Miami River, falling at a rate of about five miles per hour over the spillway (pictured). When the first water passed through the hydraulic in January 1845, it powered the flour mill and barrel-stave machinery operated by John Erwin and William Hunter, located about where the Red Trunk clothing store is at right.

Two

Papermakers

Paper mills were among the earliest manufacturing facilities that located their operations along the water resources of southwest Ohio. As many as 30 independent mills were concentrated along the Great Miami River and Miami-Erie Canal during the 40 years from 1890 to 1930; about 20 were still producing paper as recently as the 1960s. Several mills were built on the Miami-Erie Canal at Crescentville, Port Union, and Rialto. Mills in Dayton, West Carrollton, Miamisburg, Franklin, Middletown, Excello, Hamilton, and other towns were located near the Great Miami River. The most important paper mills constructed in Hamilton were those built by William Beckett and his associates, Louis Snider and his sons, and Peter G. Thomson.

William Beckett, John L. Martin, and Francis D. Rigdon formed the Miami Paper Mill in 1848 along the Hamilton hydraulic. The company went through several name changes until it was incorporated as the Beckett Paper Company in 1887. The mill began by making newsprint from rags for Cincinnati newspapers. Among the company's customers was the Cincinnati publishing company of Truman and Smith, which published the *McGuffey Readers*. Beckett made the paper for the majority of the readers printed before 1901. The company created the Buckeye Cover range of papers, and a variety of other cover, text, opaque, offset, and bond papers. Beckett Paper merged with Hammermill Paper in 1959 and then became part of the International Paper Company in 1986 and Mohawk Paper in 2005.

Louis Snider built his Franklin Paper Mill on the hydraulic in 1854 and his Fairgrove Mill on the Miami-Erie Canal in 1868. Snider's sons built two additional mills on the same lot in 1880 and 1891. These operations were sold and became the Sterling Paper Company, which left Hamilton by 1918.

Peter G. Thomson's Champion Coated Paper Company, begun in 1894, grew to be the largest coated-paper facility in the world. It was a popular Hamilton employer until International Paper, which had acquired all of Champion International's assets in 2000, sold the mill to Smart Paper in 2001.

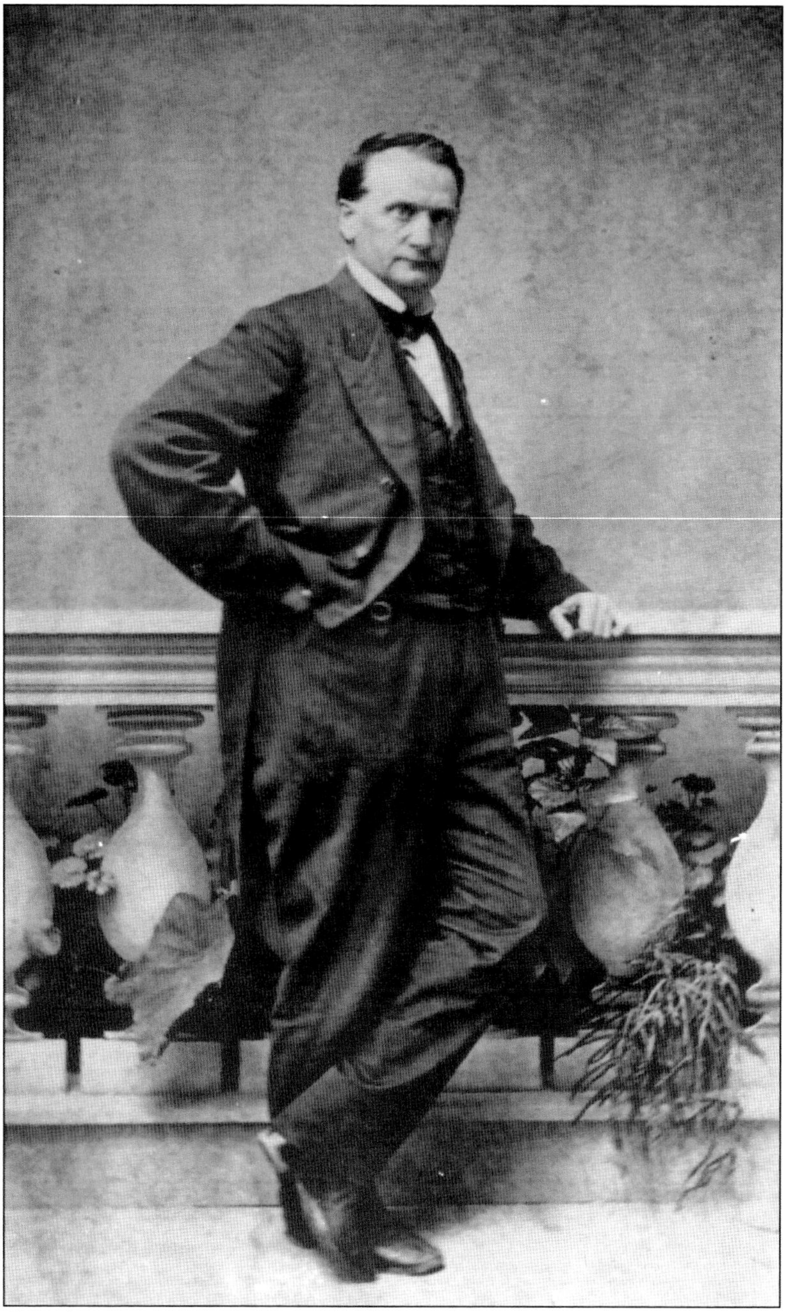

William Beckett (1821–1895) was a young lawyer and partner in a law firm with his father-in-law, John Woods, when he, Adam Laurie, John Martin, Frank Martin, and Francis Rigdon started the Miami Paper Mill along the Hamilton hydraulic in 1848. In 1854, Beckett loaned Clark Lane $1,000, with which he started Clark Lane and Company, a foundry operation that evolved first into the firm of Owens, Lane and Dyer, and then became the Hooven, Owens, Rentschler Company in 1882. The mill, after several name changes, was incorporated as the Beckett Paper Company in 1887. Beckett would lead the mill from its inception until 1876. Members of the Beckett family would manage the company for 126 years, from 1848 to 1974.

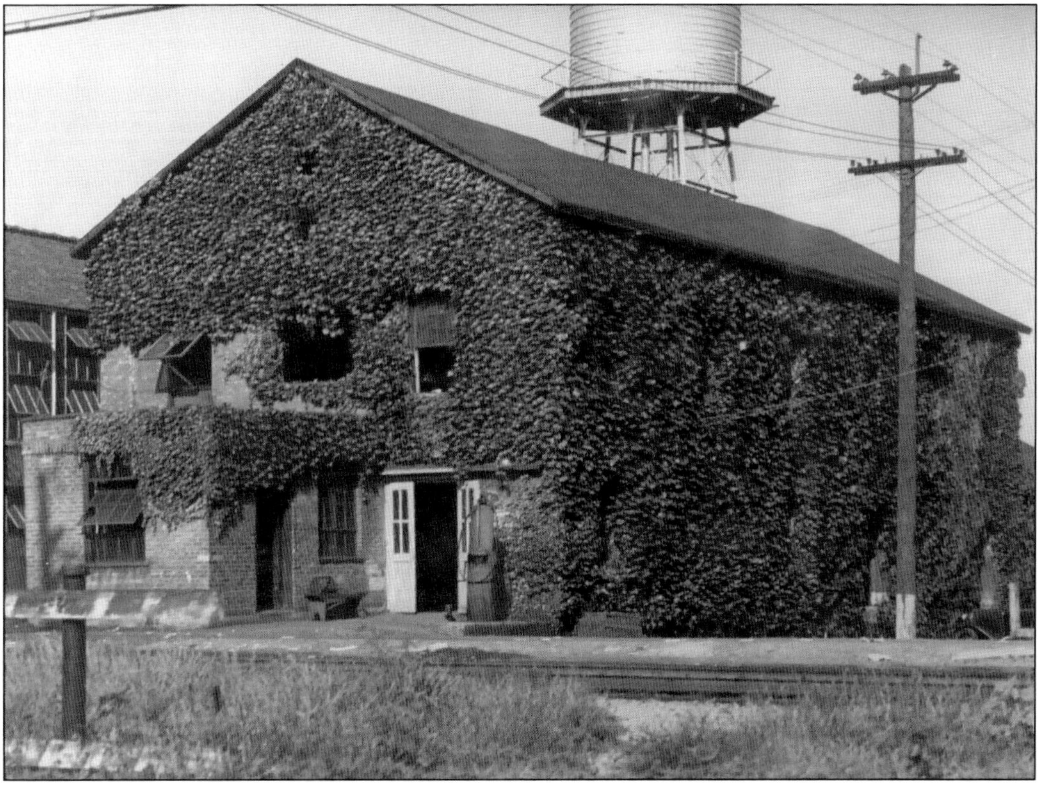

A portion of the original 1848 Miami Paper Mill is seen in this undated photograph. That first mill included two washers, two beaters, steam boilers, and a 62-inch-wide paper machine, and the total cost of the mill and its machinery was about $12,000. Early company records indicate that the highest-paid skilled operators and mechanics employed at the mill were paid $1.32 per day, and that a day consisted of 12 to 14 hours of continuous work.

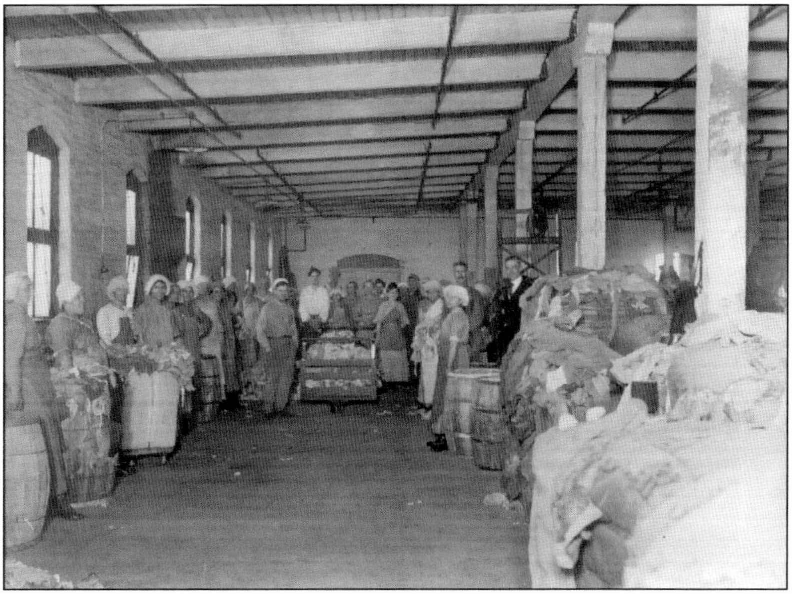

The first paper produced by Beckett Paper, in 1848, was newsprint, made entirely from rags, that was sold to Cincinnati newspapers. For many years, rag newsprint was the mill's primary product. Employees pose as rag sorters on the top floor of the mill for this photograph, taken sometime in the early 1900s.

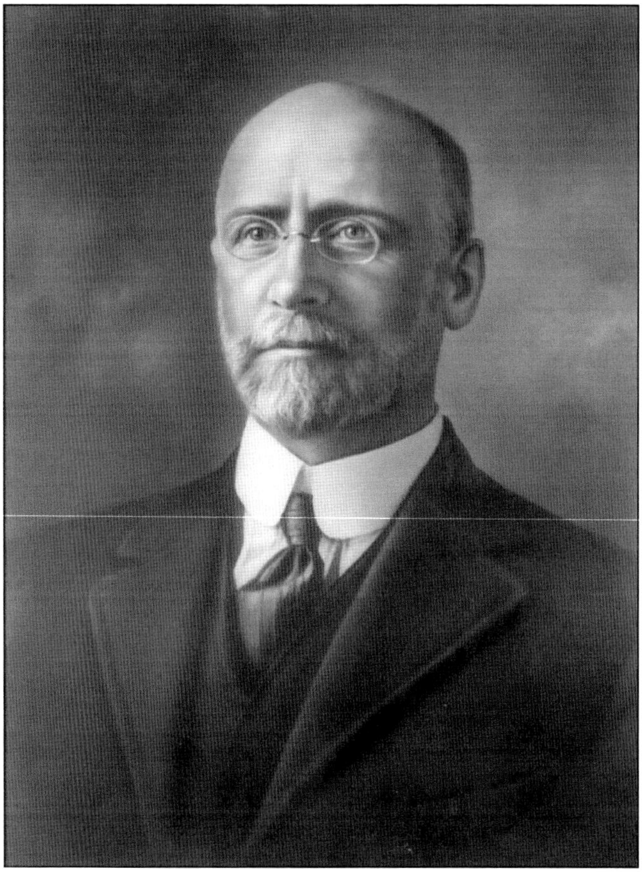

Thomas Beckett, son of Beckett Paper's founder, joined the company in 1876, took over its management in 1885, and led the firm until his death in 1923. In 1894, Beckett created and developed Buckeye Cover paper, the first full range of high-quality colored cover paper. This product line was the mill's best-known proprietary brand and was considered the industry standard for cover paper, with excellent appearance and utility.

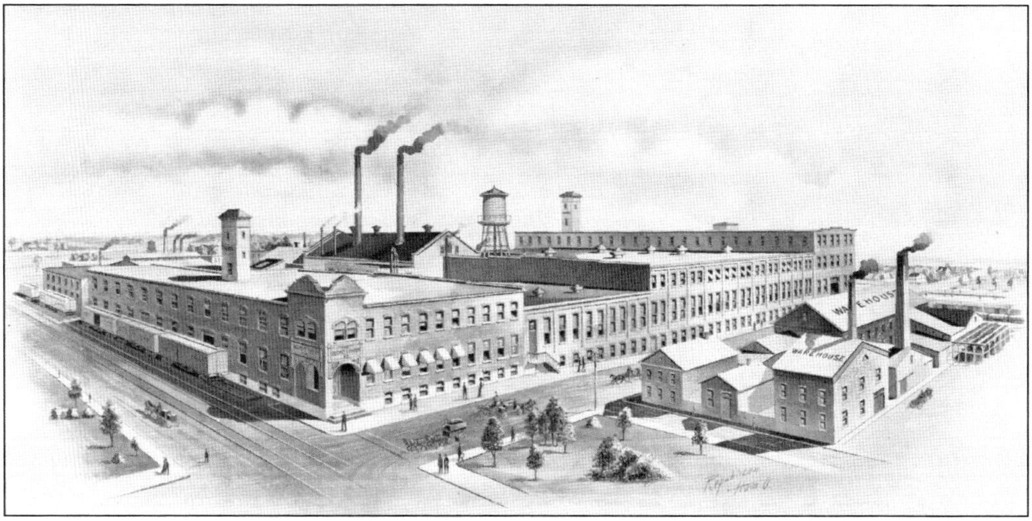

This artist's rendering of the Beckett Paper Company plant appeared in a 1914 Butler County atlas. The mill went through several expansions and modernization of machinery to create new products and meet changes in market demand. In 1958, the company added an additional 125,000 square feet of floor space when it acquired two large buildings adjacent to the mill that had been used by the Baldwin-Lima-Hamilton Corporation.

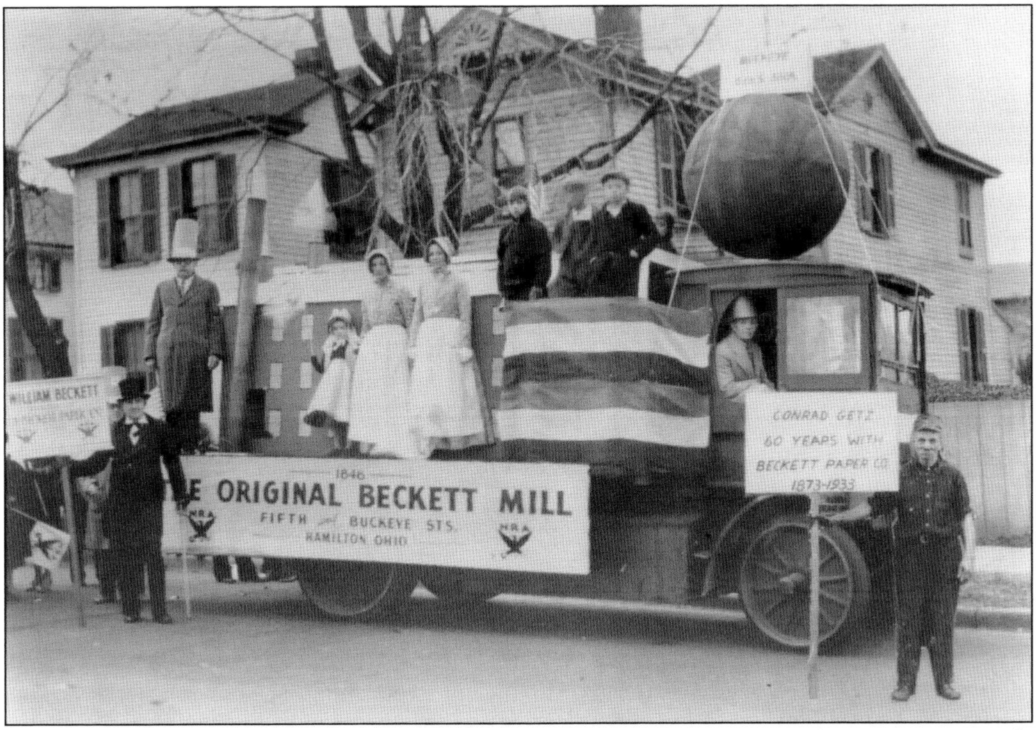

This was the Beckett Paper Company's float for a 1933 National Recovery Administration parade in Hamilton. Guy Beckett, grandson of founder William Beckett, stands in full dress at the far left. At the far right is 60-year employee Conrad Getz. Posing on the truck next to a replica of the original mill are, from left to right, Tom Warren, Mary Beckett, Hazel Wiseman, Marie Werk, Charles Barber, and two unidentified persons. The truck driver is Joe Davis.

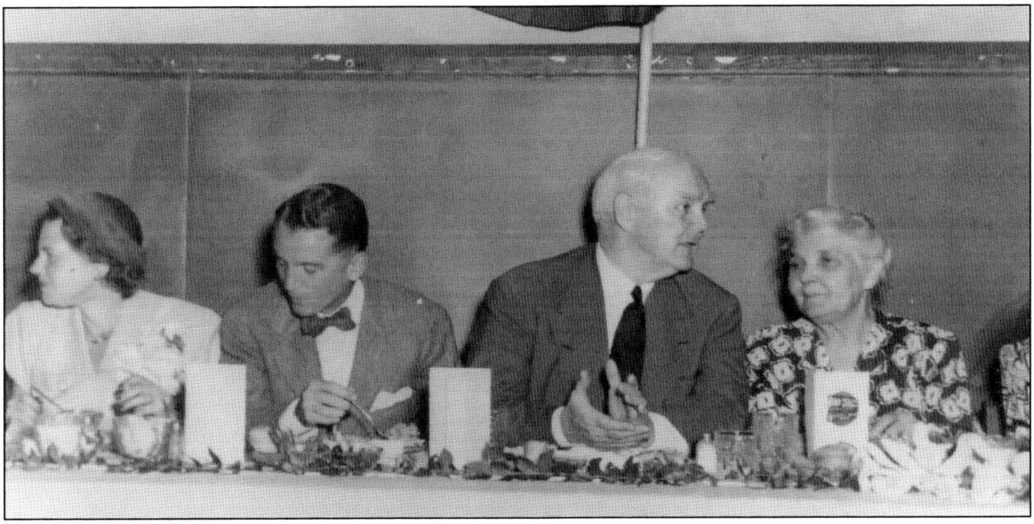

On June 8, 1948, Beckett Paper observed its 100th year of being in business with a banquet. At the event were, from left to right, Fritzi Mueller Beckett, William Beckett, Reuben B. Robertson Sr., and Mary Millikin Beckett. William Beckett would be the last member of his family to head the company. Robertson was the son-in-law of Peter G. Thomson, founder of the Champion Coated Paper Company.

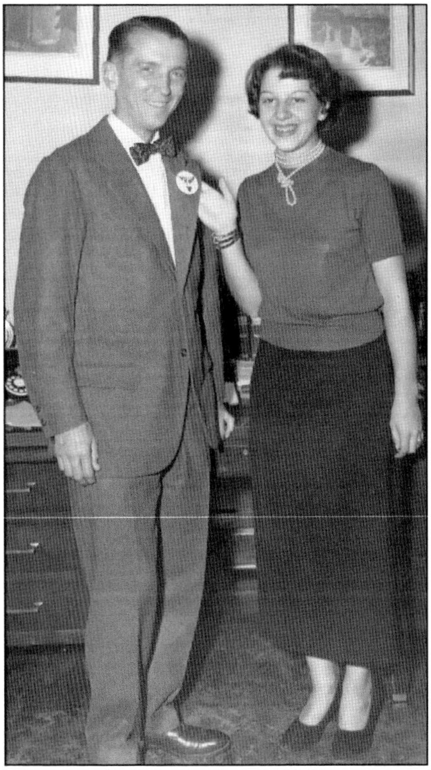

William Beckett (left), a grandson of the paper company's founder, is being made an honorary Y-Teen member by Marjorie Ernst in 1950. Beckett served six terms as a city councilman between 1940 and 1955, two terms as mayor of Hamilton, and one term as an Ohio state senator. He was president of Beckett Paper from 1958 to 1974.

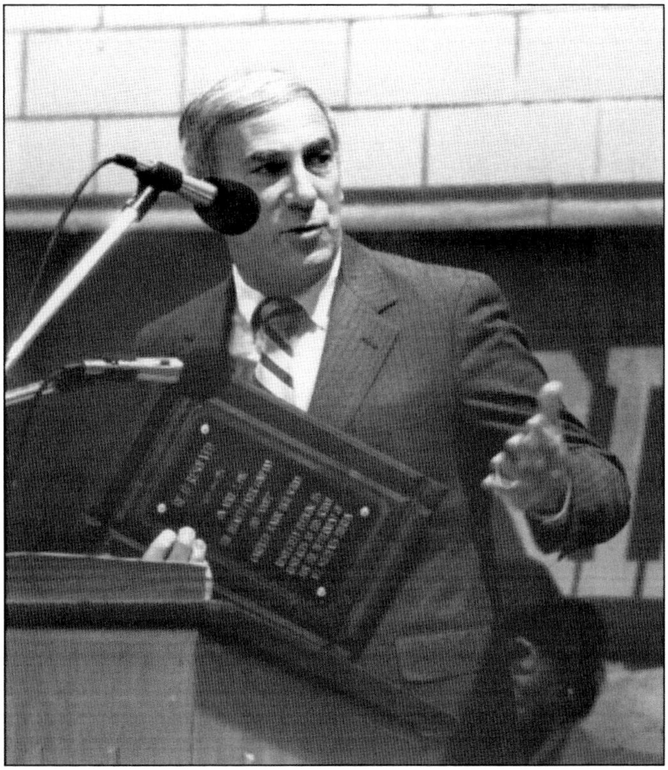

Beckett Paper president David L. Belew speaks during a ceremony marking the company's adoption of Garfield Junior High School in April 1987. Belew, who joined the firm in 1960 as manager of advertising and sales promotion, became president in 1974 and served in that position until he retired in 1992. During his tenure, the company, which had merged with the Hammermill Paper Company in 1959, became a part of International Paper Company in 1986.

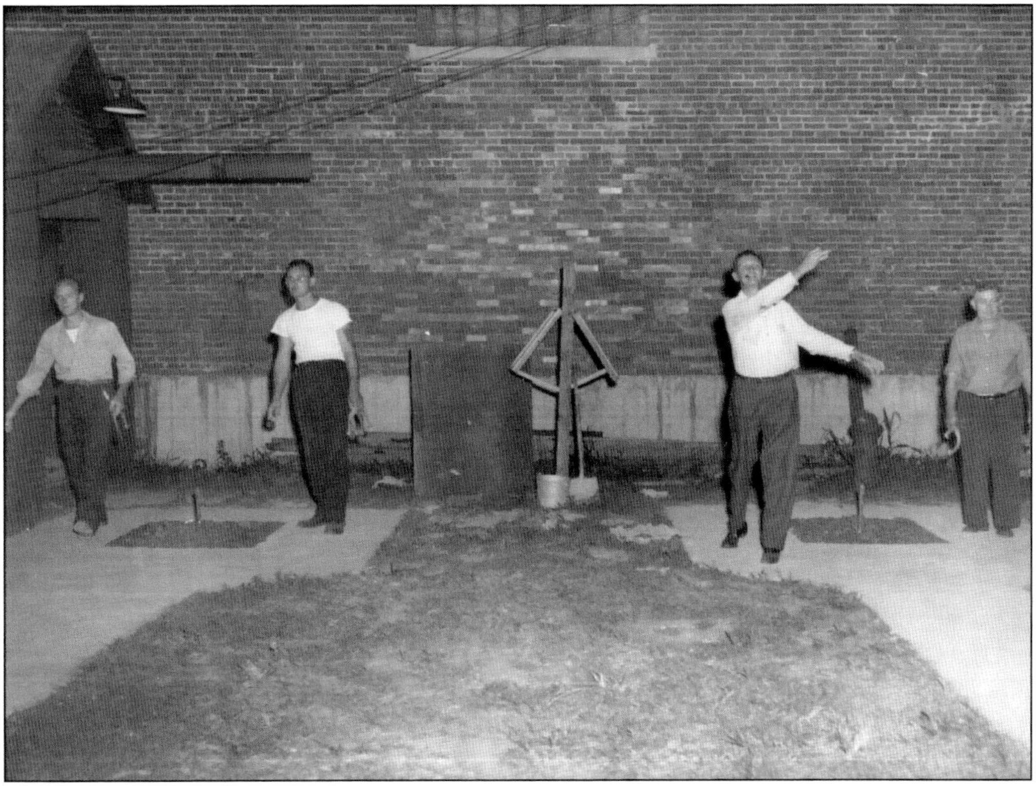

Beckett Paper supported the recreational health of its employees as much as possible. The company sponsored employee bowling teams, baseball teams, basketball teams, and a wide variety of activities. In this undated photograph, four unidentified members of the office staff take advantage of a horseshoe pitch pit near the guard shack at the mill.

The buckeye leaf served as the Beckett Paper Company's corporate logo for about 100 years, until an updated version was created in 1998. The buckeye tree, native to the Midwest, was named Ohio's official state tree in 1953, three years after Ohio State University adopted the buckeye as the school's nickname.

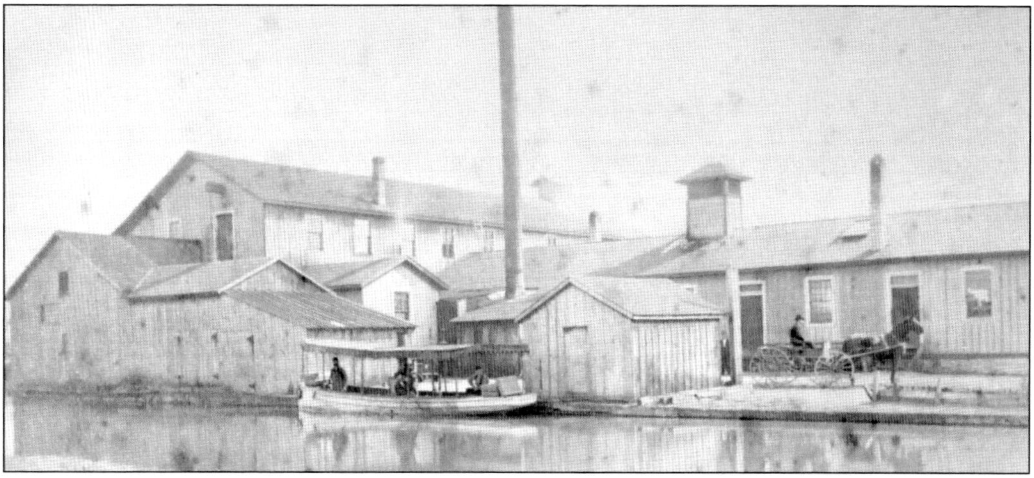

Louis Snider built the Fairgrove Paper Mill (above) in 1868 on the Miami-Erie Canal and made book paper and newsprint. He had built the Franklin Paper Mill at Buckeye and North Front Streets in 1854. The men in the photograph are, from left to right, Louis Snider, Louis Snider Jr., Henry J. Snider, William Pfau, and Edward T. Snider. After the father's death in 1875, his sons operated the mills and built the Fordham Paper Mill in 1880. They built a second mill on the same property at Second and Mill Streets in 1891. By 1899, the four Snider mills had been sold. J. Howard Friend, the buyer, kept the mills at the Second and Mill Streets location, operating as the Sterling Paper Company (below). It manufactured vegetable parchment paper, a greaseproof paper used in dairy packaging and meat packing to wrap products. After Friend's death in 1918, the company's operations moved to West Carrollton, Ohio.

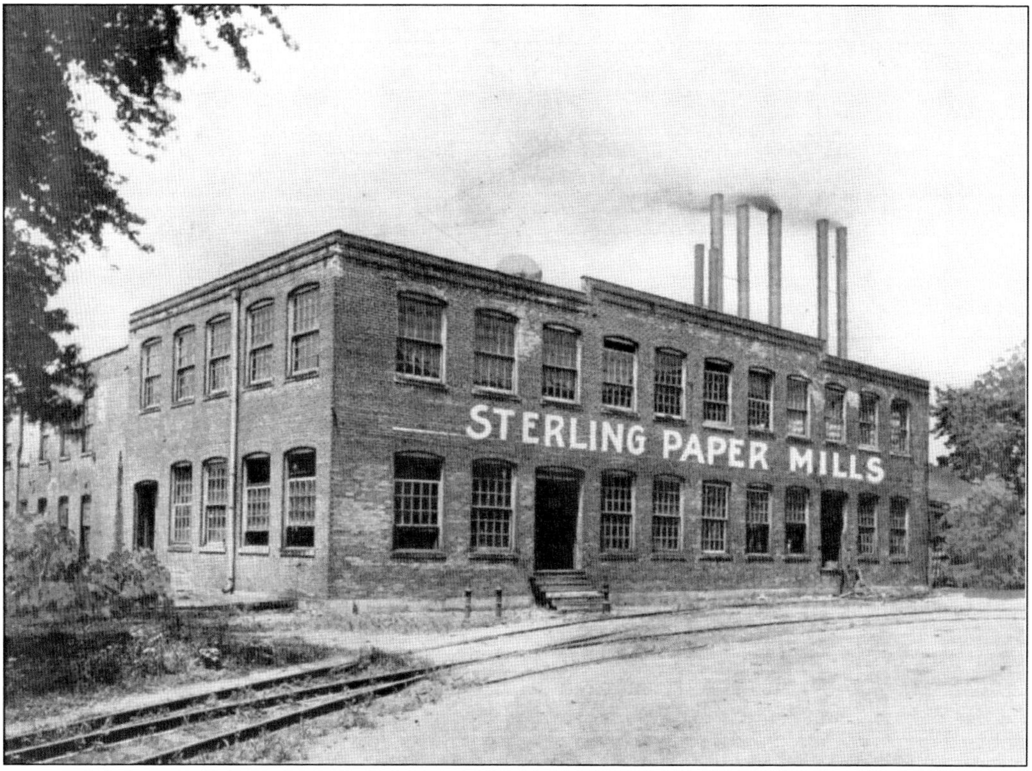

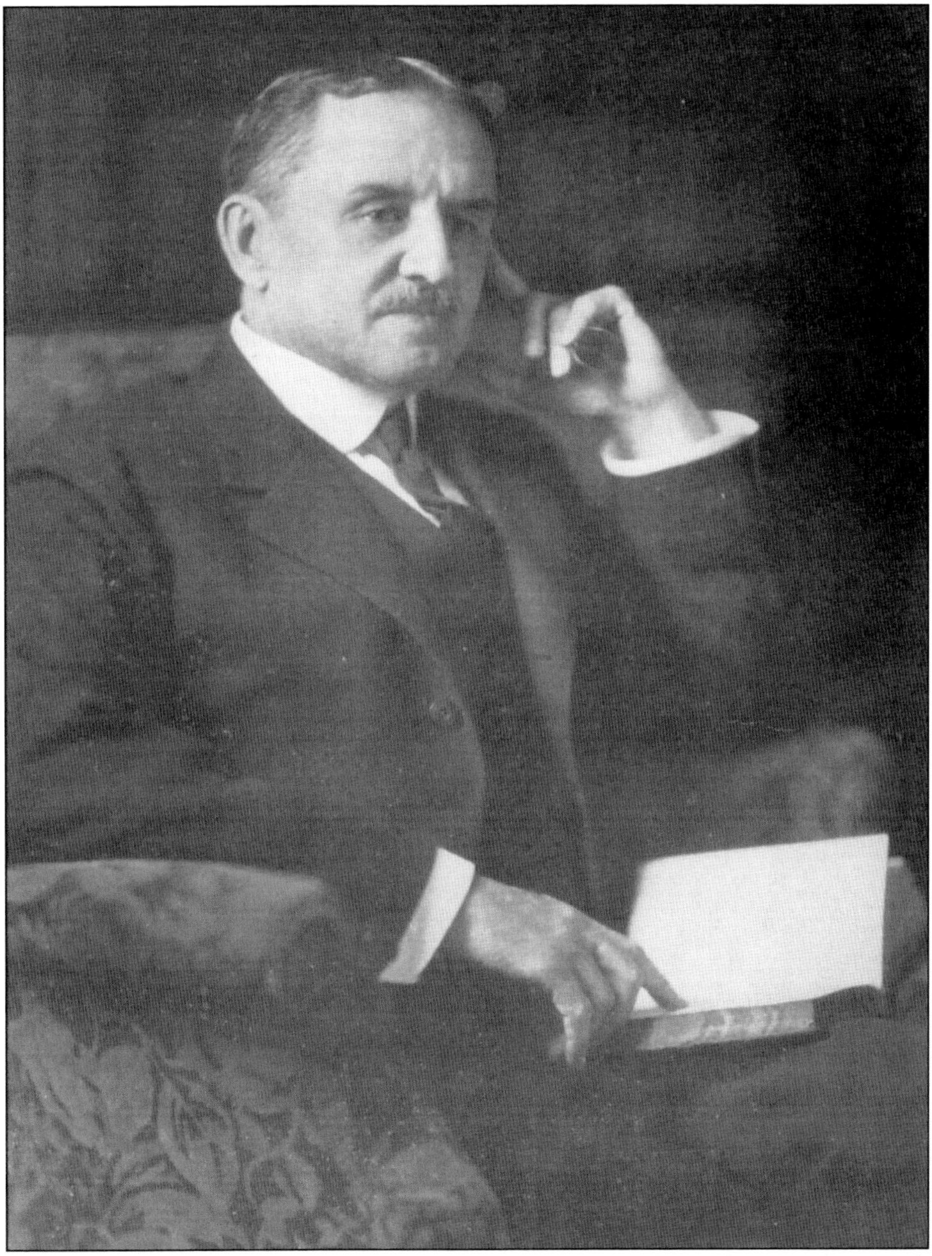

Peter G. Thomson founded Champion Coated Papers in 1893 and served as president until 1931. In 1891, Thomson bought a 200-acre tract of land in Hamilton, west of the Great Miami River, formed a real estate syndicate, and developed the area as Prospect Hill. He came up with the idea to build the Black Street Bridge to help sell homesites in his development. Thomson, who had lived in the College Hill suburb of Cincinnati since 1884, built Laurel Court, his spectacular two-story, 36-room mansion, between 1902 and 1907. At Champion, Thomson fostered good employee relations by providing workers with a company cooperative store and offering group insurance plans for them and their dependents beginning in 1917. Dr. Louis Frechtling became the first full-time industrial physician at the mill in 1919. An employee cafeteria opened at the mill in 1926, and the Chaco Credit Union was started in 1938.

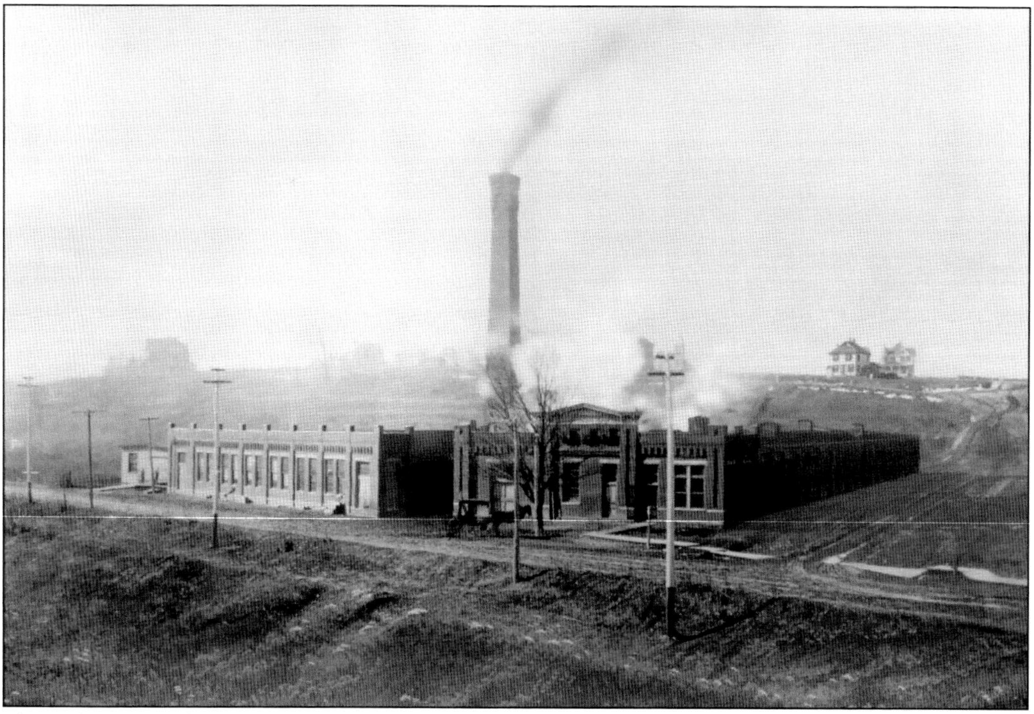

Shown here are Champion Paper's first mill, built in 1894 (right), and the company's first expansion, completed two years later (left). An 1898 flood damaged the mill, and a fire destroyed it in December 1901. By June 1902, the coating mill had been rebuilt, and the company's first papermaking mill began production. A second papermaking mill was constructed in 1907. By 1910, the Hamilton Champion mill was the largest in the world.

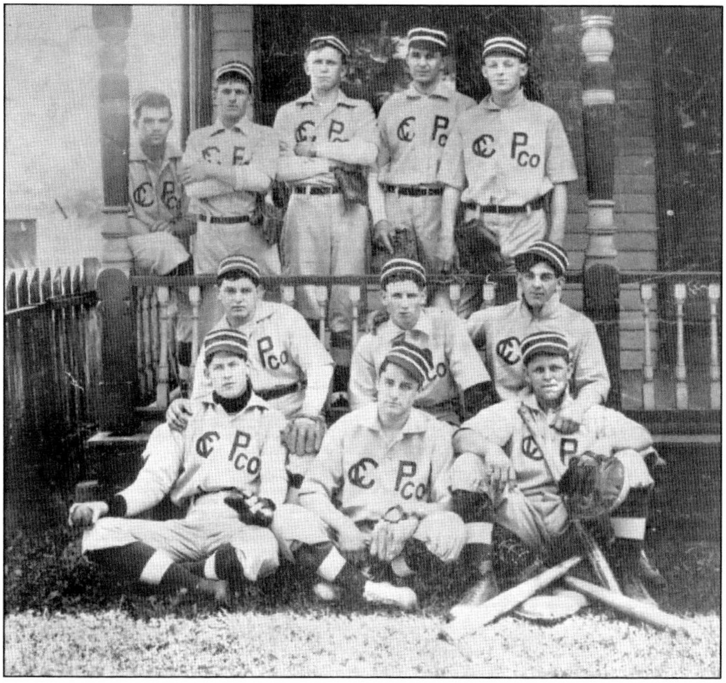

Champion Coated Paper sponsored many recreational activities for its employees. Members of the company's 1904 baseball team were, from left to right, (first row) Clarence Wheeler, Logan Thomson, and John Sampson; (second row) Joe Brunner, Harvey Sutton, and Al Myers; (third row) Claude O'Dell, Dan Shuler, George Uhlenbrock, Jim Bauman, and Ed Sanders. Thomson, son of the company's founder, became Champion's president in 1935 and served until his death in 1946.

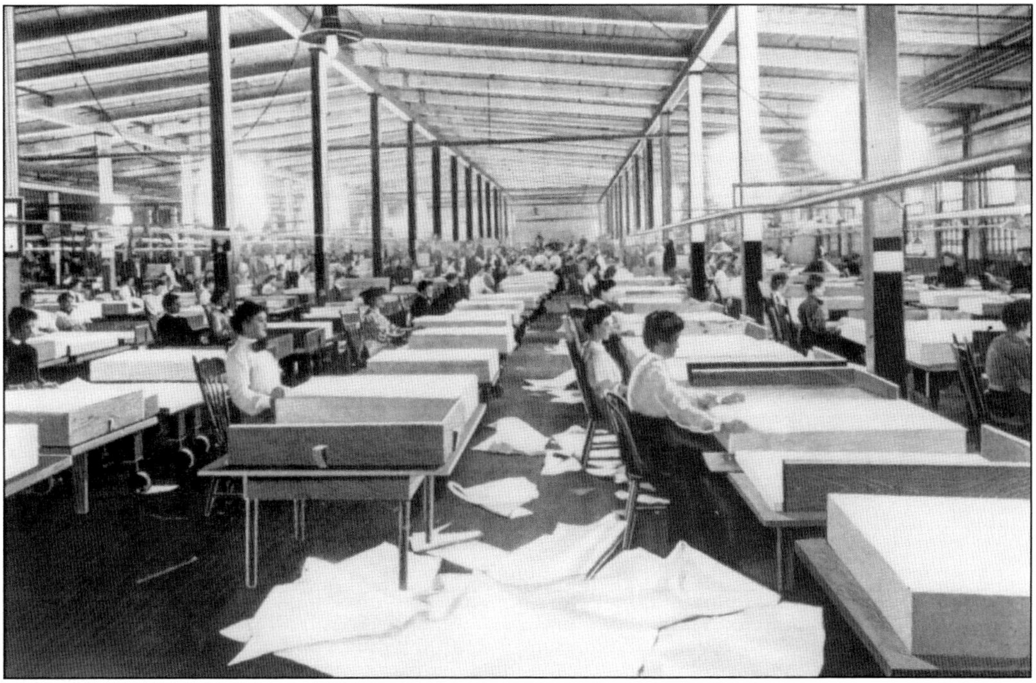

The Champion Coated Paper Company sorting and inspection room is seen here around 1905. Nearly all employees working in this room were women who inspected both sides of every sheet of coated paper and examined as much as 2,000 pounds of paper each day. This posed photograph involved special techniques to capture the activity in the room, which was 244 feet wide and 304 feet long.

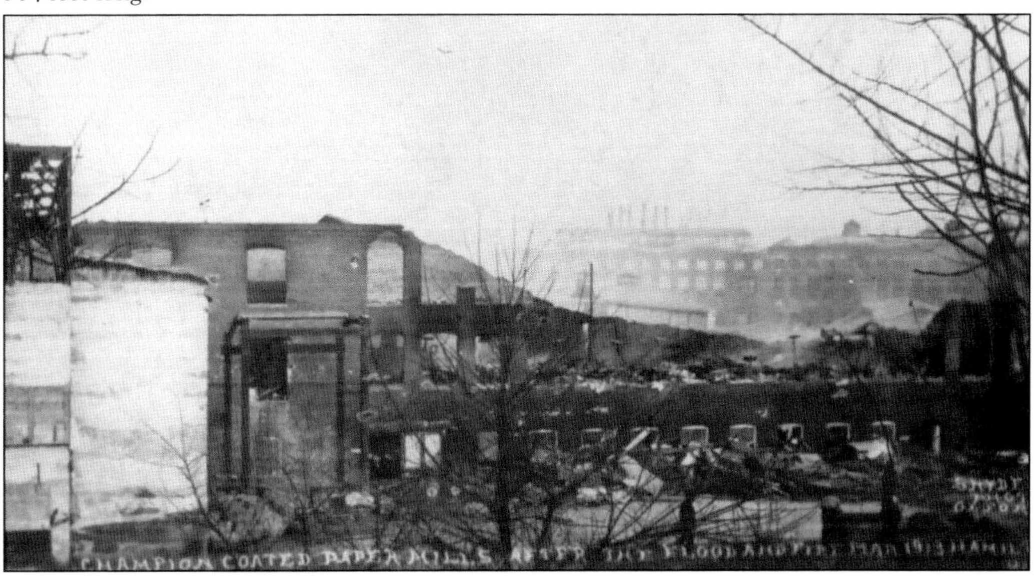

A flood damaged much of Champion Coated Paper's mill in 1898, and a 1901 fire completely leveled the facility. After both events, the company rebuilt its mills, only to have them destroyed during the Great Miami River flood and fire in March 1913. This postcard shows the damage to Champion's mills after the 1913 events. The mills were rebuilt, and production was resumed in three months.

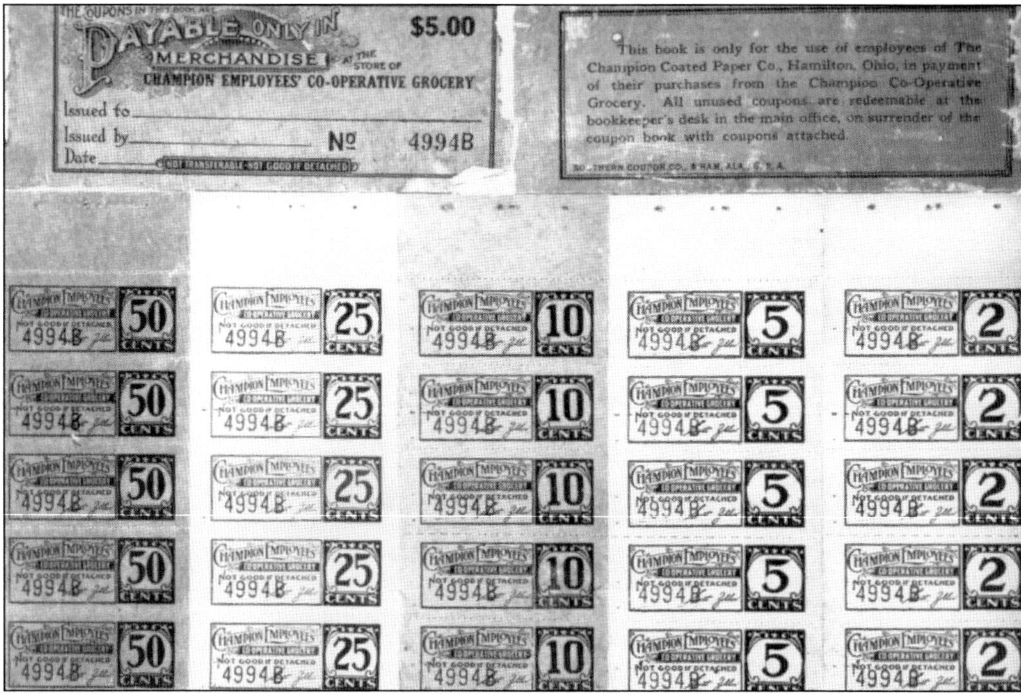

From 1917 until 1934, Champion Coated Paper employees could purchase meat, groceries, and other household needs at discounted prices at the company cooperative store. Workers bought these coupons and used them to make purchases at the store. Since many of the groceries were sold at wholesale prices, company scrip offered a good deal for employees.

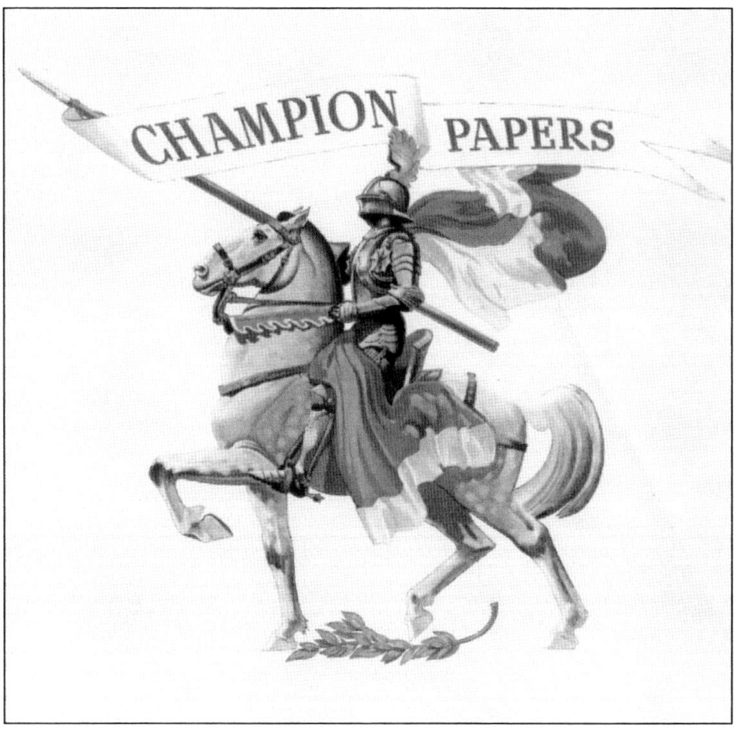

The Champion Coated Paper knight logo made its first appearance in October 1925, shortly after the company added an advertising department. The logo, featuring a mounted medieval knight in full armor with a flowing robe and carrying a lance, was used to label products made at the Hamilton mill. It was also used on internal company publications and in advertising to customers. The image gradually disappeared from use as a result of name changes and mergers.

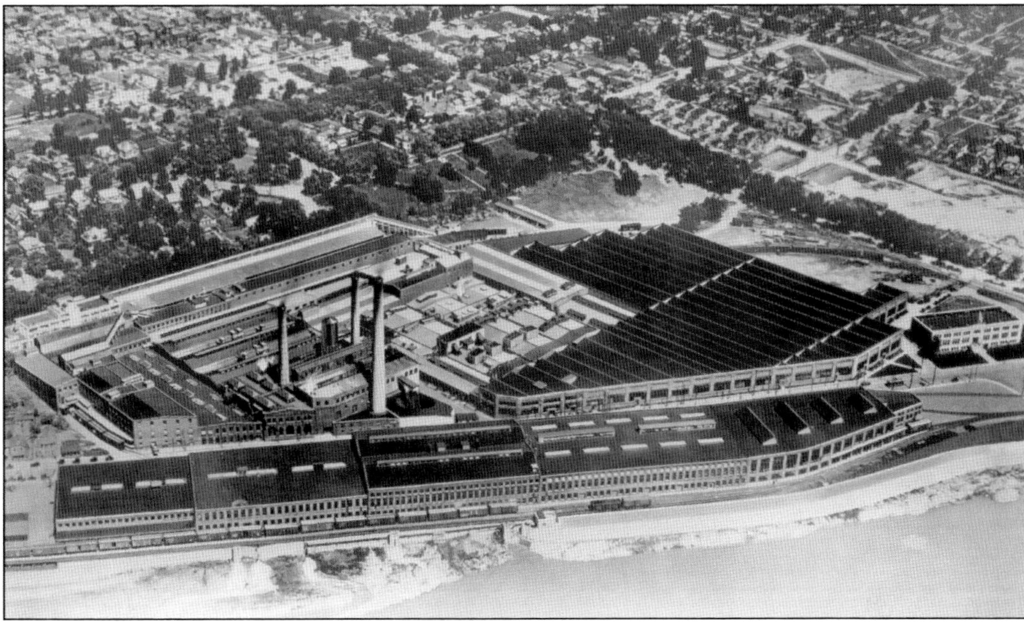

Champion's facilities, rebuilt after the 1913 flood, were then expanded, with the addition of coating machines, modernization of equipment, and construction of additional buildings. This aerial photograph of the mill was taken in 1928. At the start of the Great Depression, more than 4,000 people worked at the facility. The number employed dropped to about 800 when Champion and all of its property was acquired by International Paper in 2000.

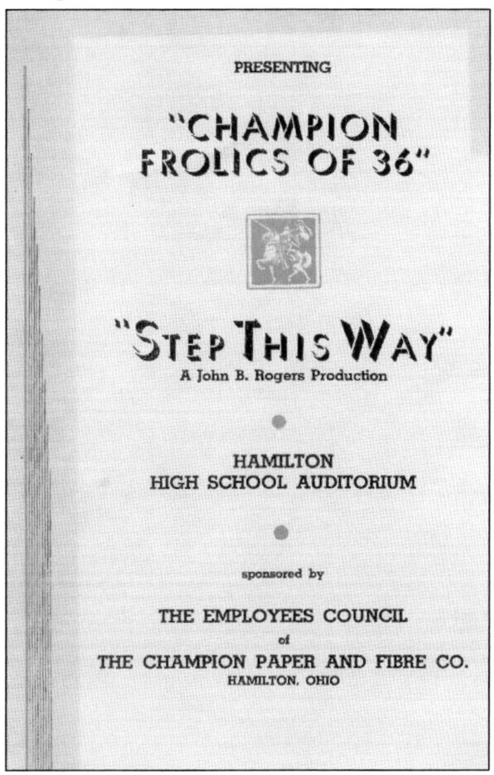

PRESENTING

"CHAMPION FROLICS OF 36"

"STEP THIS WAY"

A John B. Rogers Production

HAMILTON HIGH SCHOOL AUDITORIUM

sponsored by

THE EMPLOYEES COUNCIL of THE CHAMPION PAPER AND FIBRE CO.

HAMILTON, OHIO

Step This Way, a musical comedy, was produced in 1936 and featured 104 Champion Paper employees in the cast. There were 24 speaking roles, several chorus acts, and dance ensembles in the program. The lighting for the show was put together by Champion's electric department, and the scenery was built in the company's carpenter shop. The show was the second annual stage production put on by Champion employees.

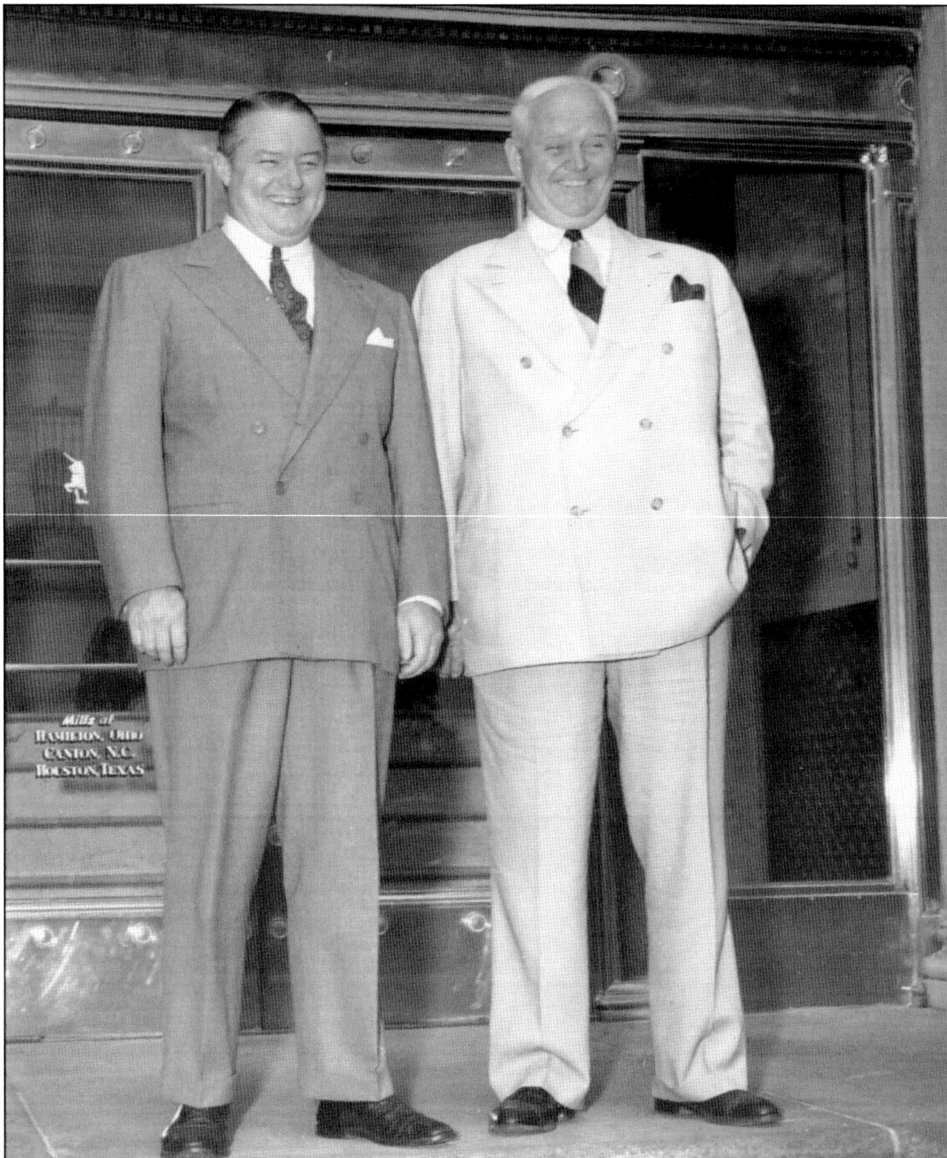

Standing in front of the company's office in 1950 are two generations of Champion Paper's leaders. Reuben B. Robertson Sr. (right) was the son-in-law of company founder Peter Thomson. In 1906, he founded the Champion Fibre Company in Canton, North Carolina, to supply the Hamilton mills with wood pulp with which to make paper. Robertson, as president, developed the North Carolina plant, guided its substantial forestry operations, and worked to create the Great Smoky Mountains National Park, which included 90,000 acres of Champion land that it sold for $3 million. He served as president of the North Carolina operation from 1907 until 1946, the year he became president of the entire company. Reuben B. Robertson Jr. (left) joined the company in 1930 and advanced through several executive positions until he became Champion president in 1950, the same year his father was elected chairman of the board. Reuben Jr. was named Man of the Year for 1953 by the National Association of Manufacturers and appeared on the cover of *Business Week* magazine. In 1960, Robertson Jr. was killed in a traffic accident.

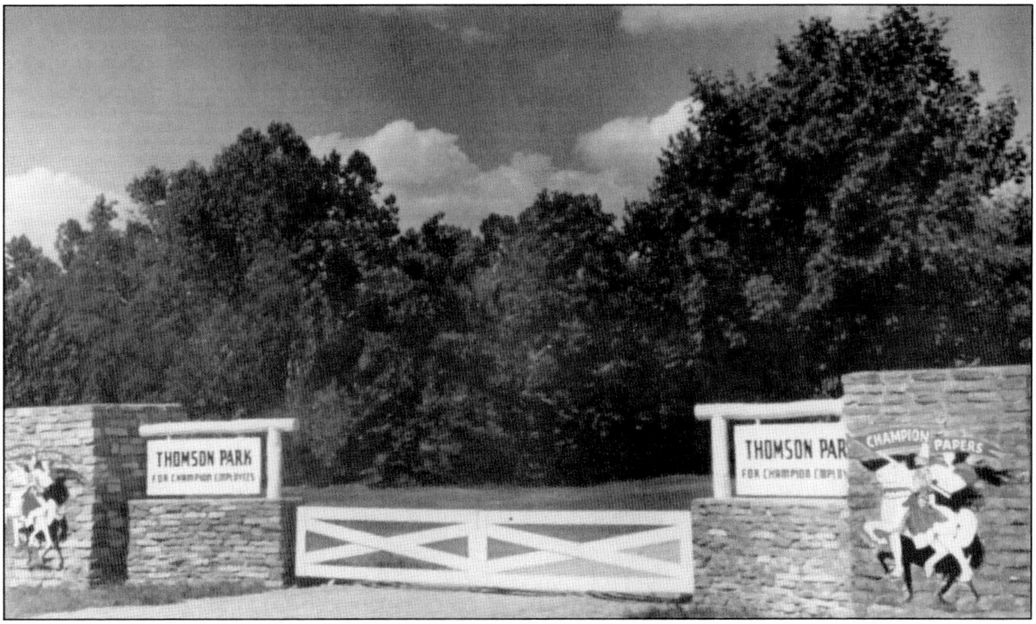

In 1937, Champion Paper leased six acres west of Hamilton for use as a campsite, named Camp Chapaco, for Boy Scouts and Girl Scouts. The company bought the land in 1947 for an outdoor recreation and social center for employees and their families, renaming it Thomson Park. By 1955, the facility had been expanded to 52.5 acres. The company closed the facility in 1995 and sold it in 1997.

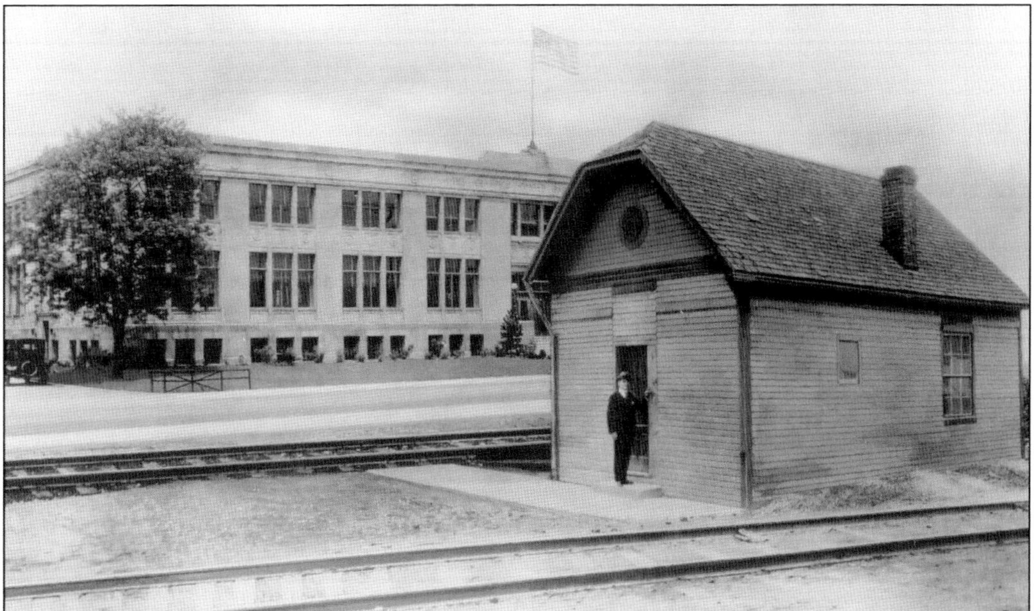

Champion Paper's first two office buildings are seen in this unique photograph. The man standing in the doorway of the company's first office building, erected in 1893, is founder Peter G. Thomson. This small building served as the company office until it was replaced in 1899. The large office building in the background, first occupied in 1925, served the company until 1961.

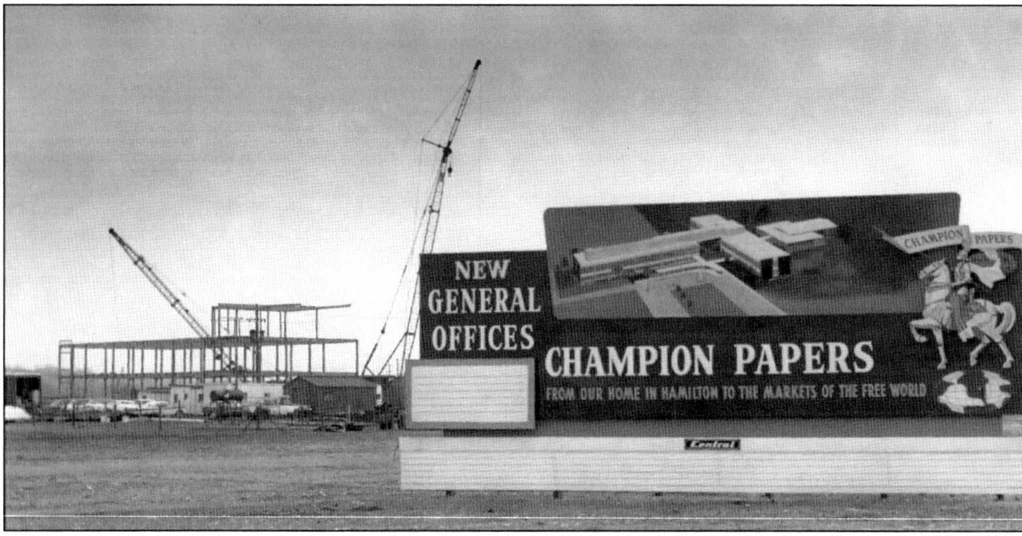

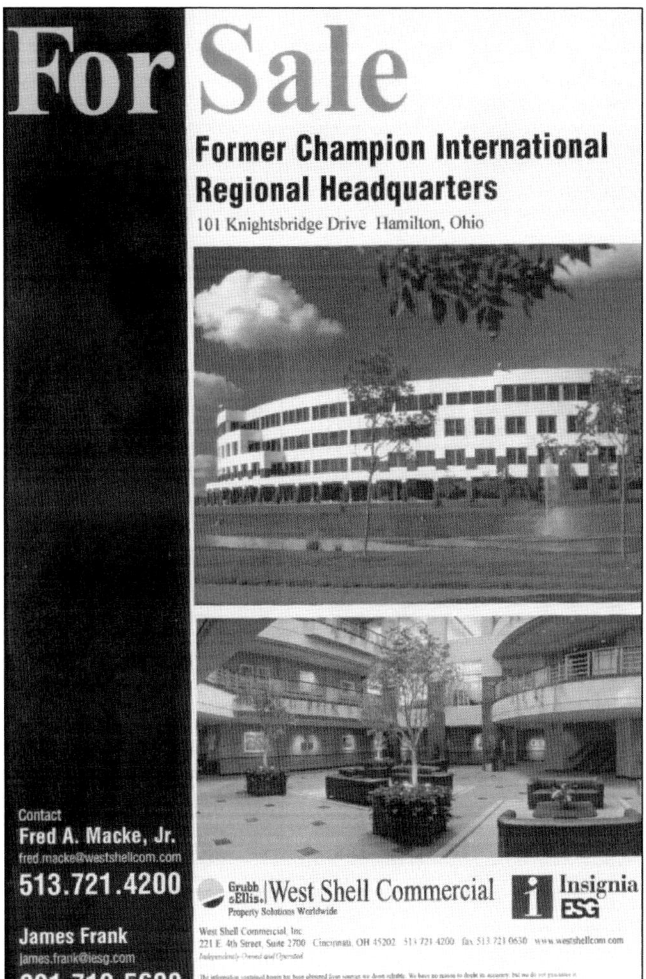

Construction of the new Champion Coated Paper administrative office building was under way when the above photograph was taken in 1960. The company moved several office functions into the building in August 1961. The facility was named Knightsbridge, in keeping with the company's knight logo. Hamilton's city council renamed South Street, the street on which the building was located, as Knightsbridge Drive in appreciation of the firm's decision to remain in Hamilton. In 1967, Champion merged with the US Plywood Corporation, and many local corporate officials moved to New York City and then to Stamford, Connecticut. The Knightsbridge office facility was expanded in 1991, but by 1997, the company announced plans to sell off its Hamilton properties. The "For Sale" brochure shown left was used to help sell the building beginning in 2000. A sale was finally consummated in 2004, when Harry T. Wilks bought the office complex for $3.5 million.

Three

Safes and Vaults

The first safe manufacturer in Hamilton was the MacNeale & Urban Safe and Lock Company, an operation that relocated to the city's west side in 1890 after doing business in Cincinnati for 43 years. Conflict among the owners caused the firm's officers to place it in receivership in 1903 and closed the company down.

By the time the Mosler Safe Company moved to Hamilton from Cincinnati in 1891, its safes and vaults were sold in many countries of the world. Business leaders in the Hamilton Improvement Syndicate persuaded the company to relocate its factory to the city. Mosler Safe's products were known for their strength and were considered burglar proof, fireproof, and capable of withstanding great pressure, including an atomic bomb explosion. Mosler safes and vault doors protected the assets in Federal Reserve Banks, the Fort Knox gold depository, nuclear missile silos, and Tennessee Valley Authority power plants. For many years, a Mosler vault housed America's Declaration of Independence, the US Constitution, and the Bill of Rights. At its peak, the company employed more than 1,000 people in Hamilton and was an anchor of the local economy for decades. Mosler family ownership of the company ended in 1967, when the firm was sold. The plant, once the largest safe-manufacturing factory in the world, ceased production in Hamilton in 1986.

Herring-Hall-Marvin Safe Company was an 1892 consolidation of four safe makers located in New York, Philadelphia, and Cincinnati. The Hamilton Improvement Syndicate enticed the firm to move to Hamilton in 1897. In 1905, the company was reorganized as a New York corporation, but it kept its works on the east side of town. Herring-Hall-Marvin remained an important Hamilton business for more than six decades, until the sale of the company was finalized in 1961. In 1991, the Herring-Hall-Marvin operation closed down, and safe-making in Hamilton was a thing of the past.

Hamilton's nickname as the "Safe Capital of the World" was well earned. A 1938 book written by the Federal Writer's Project for the Works Progress Administration claimed the total output of Mosler Safe and Herring-Hall-Marvin was about 50 percent of the world's safes and vaults.

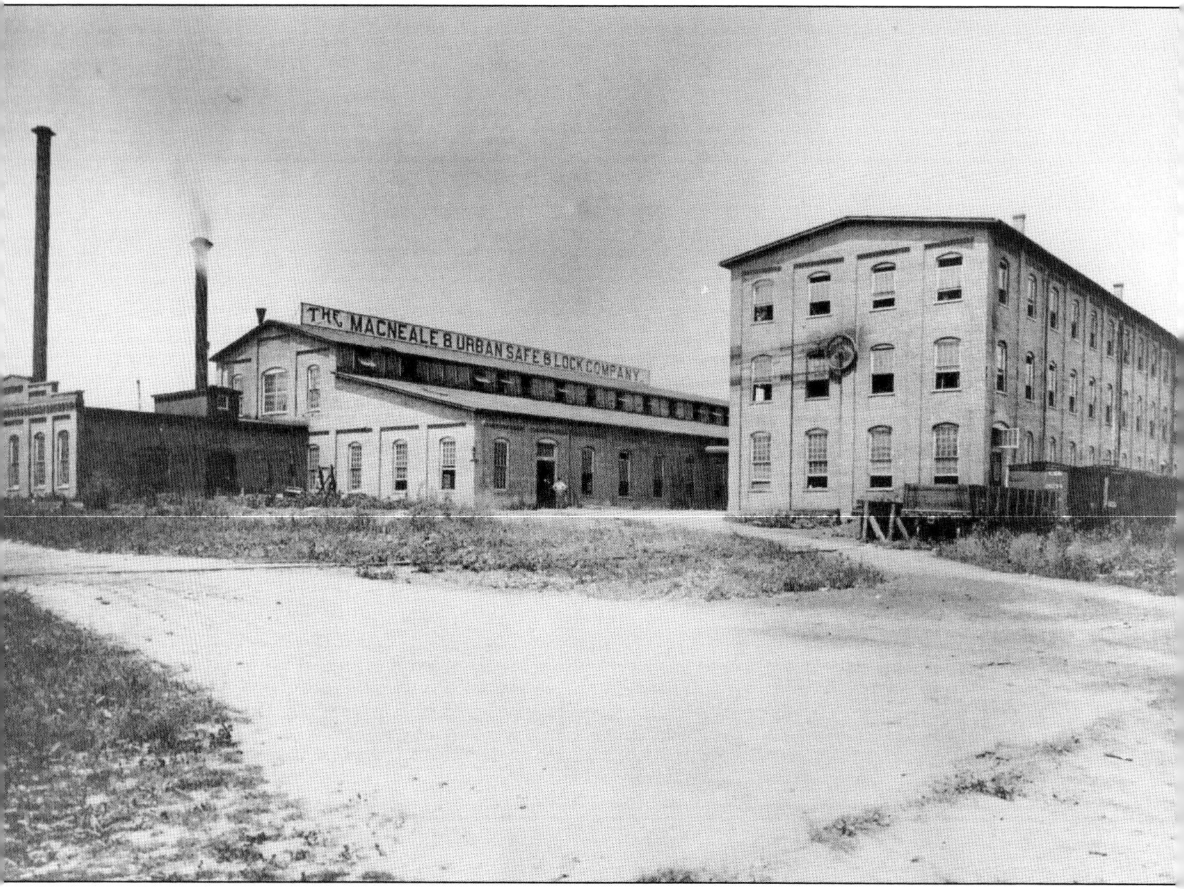

This factory housed the operations of Hamilton's first safe-manufacturing company. MacNeale & Urban began making strong boxes and key locks in Cincinnati in 1847 as the C. Urban Company. In 1889, the firm decided to relocate to Hamilton when a group of local residents raised $50,000 to buy land located at Millville and Edgewood Avenues for a new factory. Safe production began in the new plant in 1890, one year before the firm was incorporated as the MacNeale & Urban Safe and Lock Company. The company's safes, constructed with multiple layers of steel and iron welded together, were considered burglar proof and fireproof and featured combination locks. In 1903, the company was placed in receivership, and production ceased. Mosler Safe and Lock purchased the MacNeale & Urban factory in 1907. The building was later used by several manufacturers, including the Carriage Woodwork Company, Long and Allstatter, Champion Paper and Fibre Company, and the Brown Furniture Company.

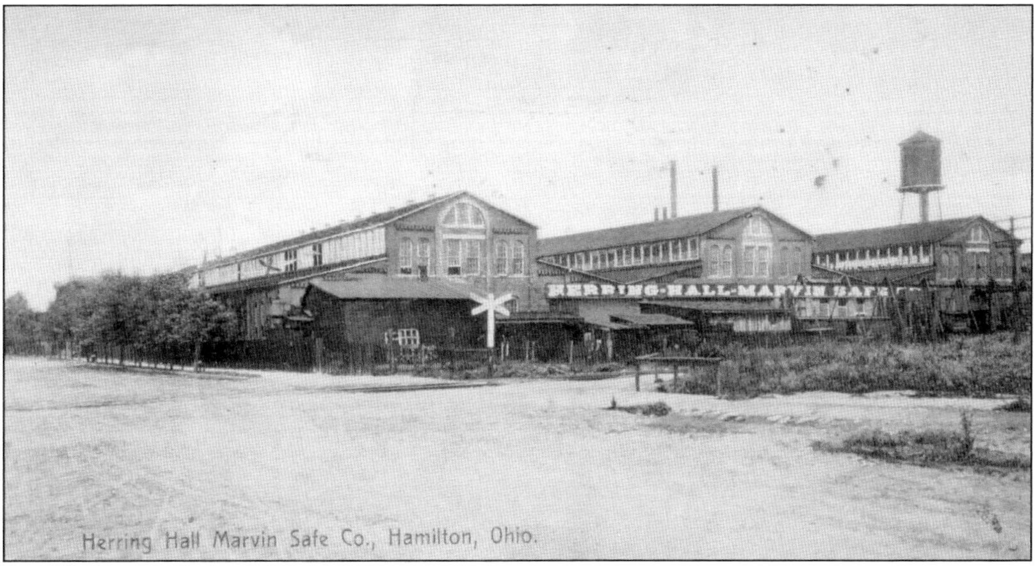

This postcard from around 1905 shows the factory of the Herring-Hall-Marvin Safe Company that was built along Grand Avenue in 1897. According to one account, the company's employees working here in 1904 received pay for 60 hours for working just 56.75 hours. At the time, the firm was considered to be treating its workers in very liberal ways. (Courtesy of Rob Wile.)

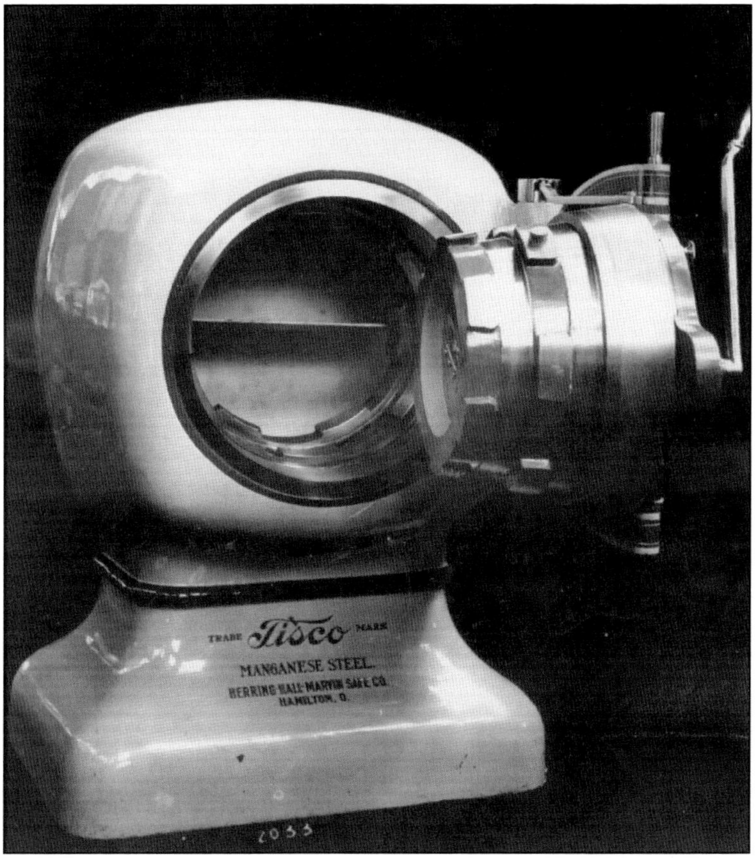

Herring-Hall-Marvin manufactured this solid manganese steel safe sometime after 1898 using a patented method of casting the steel developed by the Taylor Iron and Steel Company of High Bridge, New Jersey. The safe, with its "Tisco" steel body and 12-inch-thick door, was claimed to have double the locking strength of any other burglar-proof safe on the market. It was a popular solution to the security needs of banks that could not afford a full bank vault.

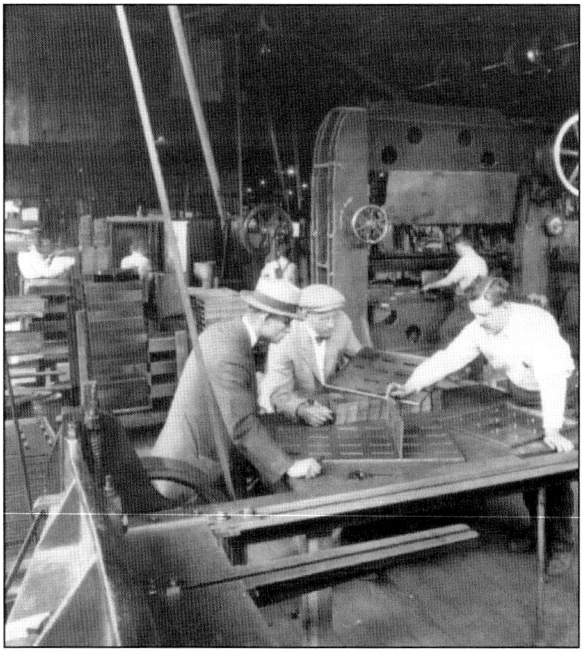

Here, three Herring-Hall-Marvin staff members study the inner workings of a fire-resistant safe in 1915. The men looking at the apparatus on the table are, from left to right, Elmer E. Watson, unidentified, and Paul S. Millice. Watson, a company supervisor, held patents for interchangeable key locks and combination-lock mechanisms. Millice, also an inventor, held patents for fire-resistant safes, temporary safes, file cabinets, and other items.

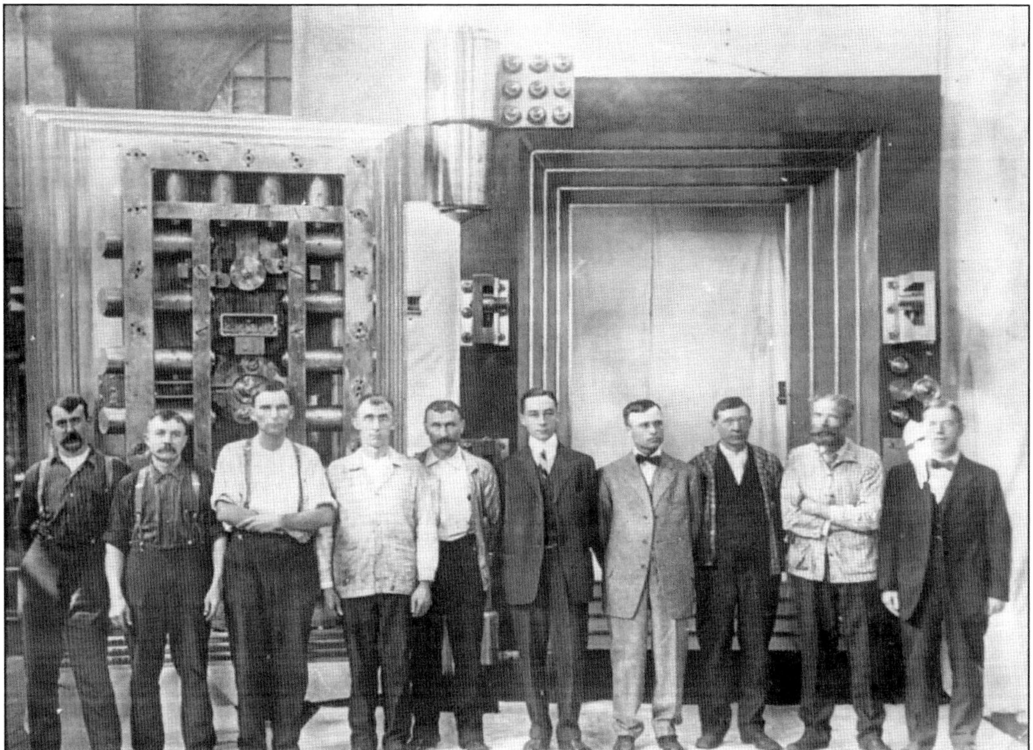

Several employees of the Herring-Hall-Marvin Safe Company stand outside the massive door and vault manufactured at the Hamilton plant in 1918. The men are, from left to right, George Schrage, Gus Jenkner, Fred Lampke, August Biermann, Henry Holstein, William F. Forepaugh (president), Elmer E. Watson (supervisor), William Schultz, Ferdinand Aker, and Walter C. Fuller.

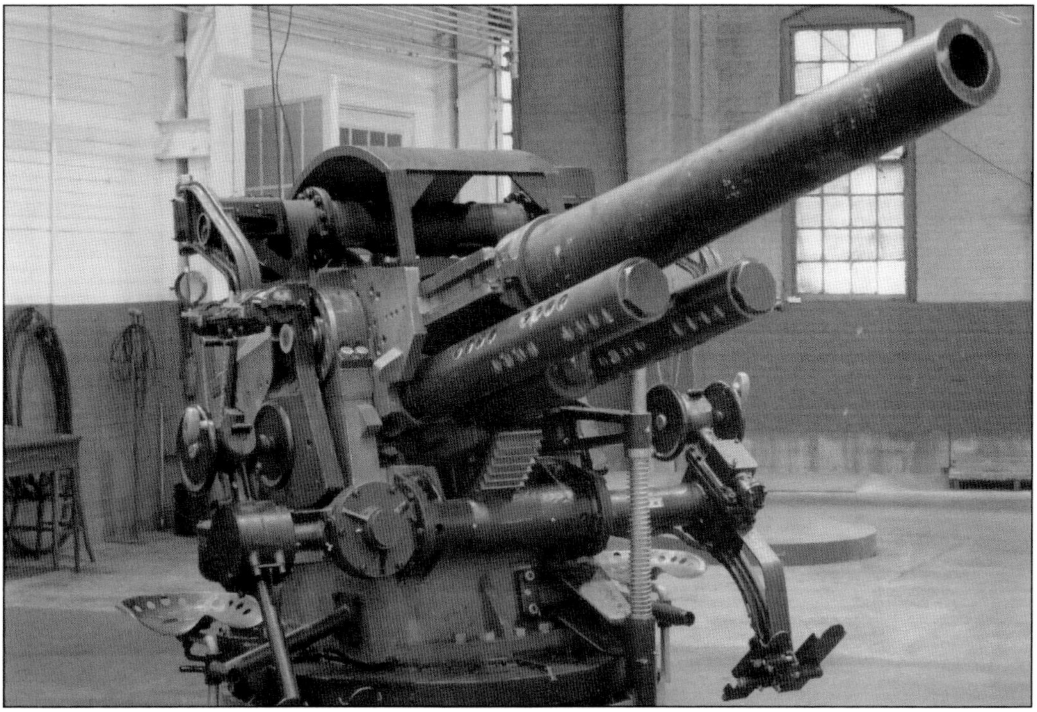

In 1943, Herring-Hall-Marvin Safe received a $3 million contract to manufacture a variety of weapons, mortar shells, and other ordnance for the US Navy. The gun shown above is one of 126 that the company manufactured for mounting on all US submarines. The design of the five-inch, .25-caliber deck gun was the result of a collaborative effort of company engineers and naval authorities. As a standard deck gun, it stayed on the submarine's deck when it submerged and provided substantial firepower for surface fighting and bombardments of enemy shore installations. The large guns shown below are three of the five-inch, .38-caliber, all-purpose antiaircraft guns and mounts produced by the company for and used exclusively on aircraft carriers of the Essex class. These vessels were considered the backbone of the Navy's combat strength during World War II.

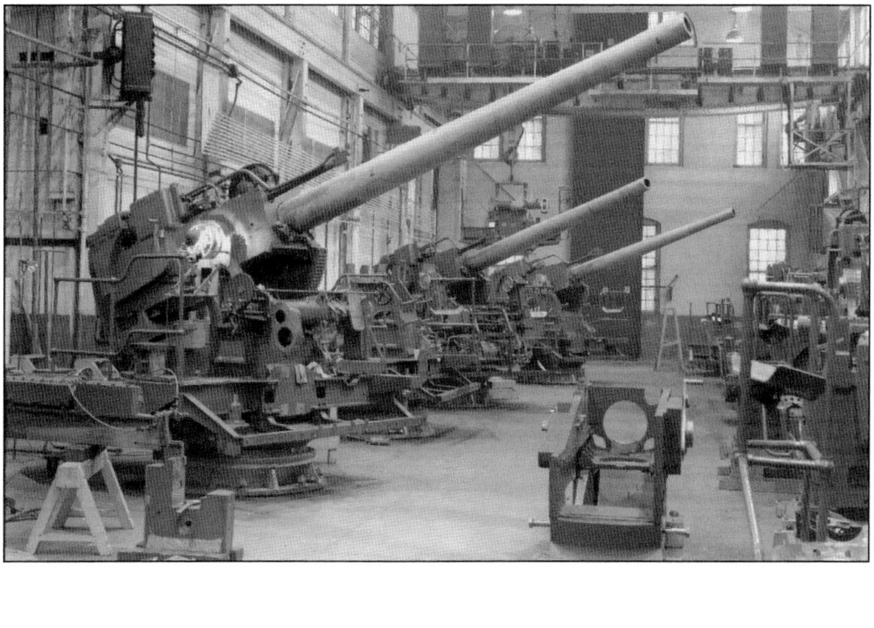

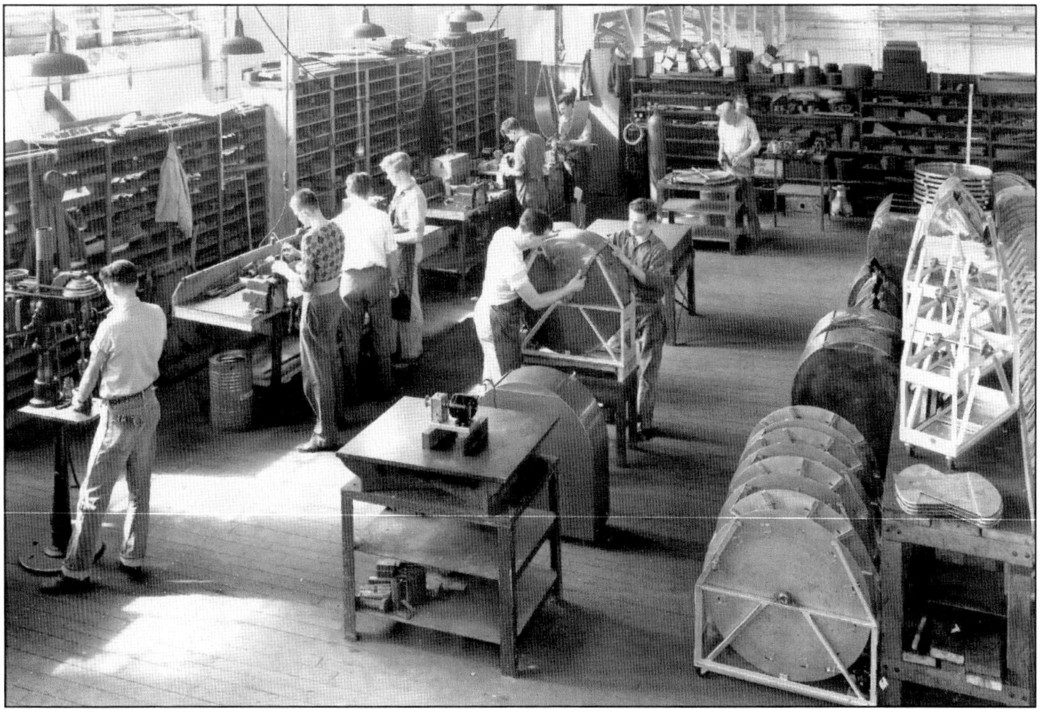

Due to changing business and economic conditions, Herring-Hall-Marvin diversified its products by developing new security and storage solutions for a wider client base. These employees are building rotary file systems designed to hold record cards for a customer. Other new products included office equipment, cashier safes, teller counter security systems, and record-protection and storage cabinets.

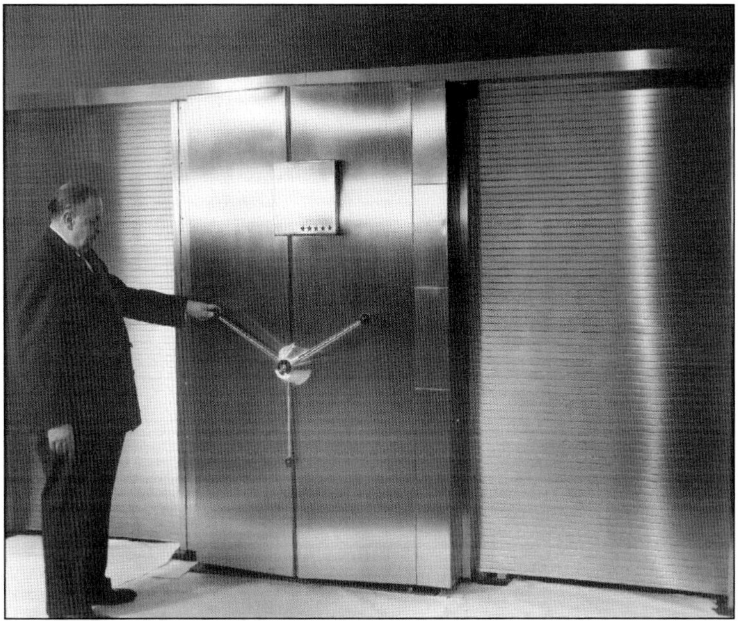

In December 1956, Warren Mosman, president of Herring-Hall-Marvin, introduced the firm's new vault door. The Five-Star Constellation door for bank safe-deposit vaults opened a full 180 degrees, had push-button control for withdrawing its locking bolts, and provided 99 percent pure electrolytic copper protection for the time lock. In less than three years, the company and its products were purchased by Diebold Incorporated of Canton, Ohio.

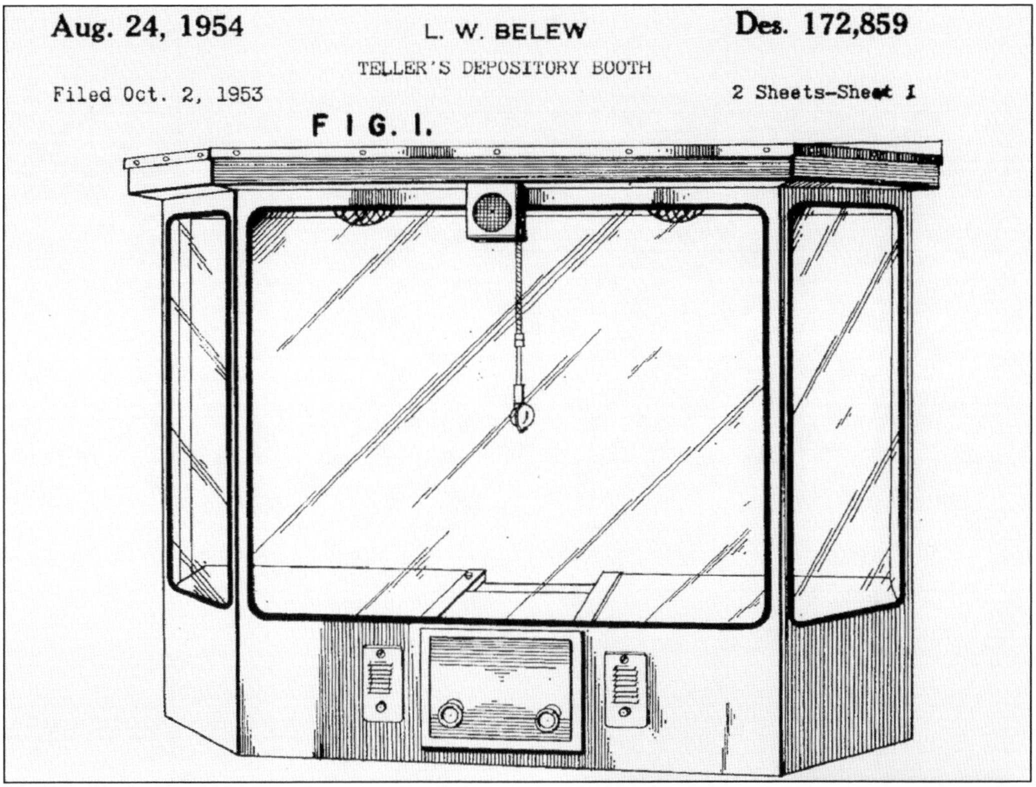

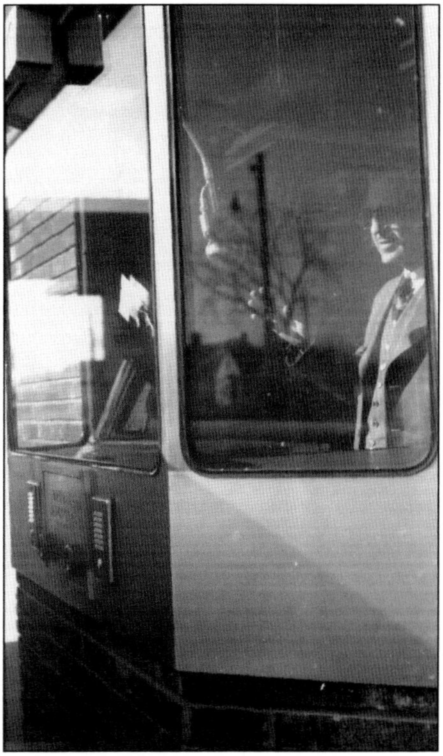

Herring-Hall-Marvin also offered its bank customers high-level security systems, such as the teller's depository booth seen above in a frontal view. The image is from patent information filed with the US Patent and Trademark Office in 1953. This "angular bay" drive-through bank window was designed by Leland W. Belew, chief engineer. Belew held several patents for other Herring-Hall-Marvin products, including the company's autobank drive-in and walk-up teller windows, rotary records file systems, vault doors, filing systems, and other products. Belew is seen in the right photograph, waving from inside one of his patented teller booths. His son David became an executive at the Beckett Paper Company and served as its president for 18 years, from 1974 to 1992. Diebold Incorporated, Herring-Hall-Marvin's parent company, closed the plant in 1991.

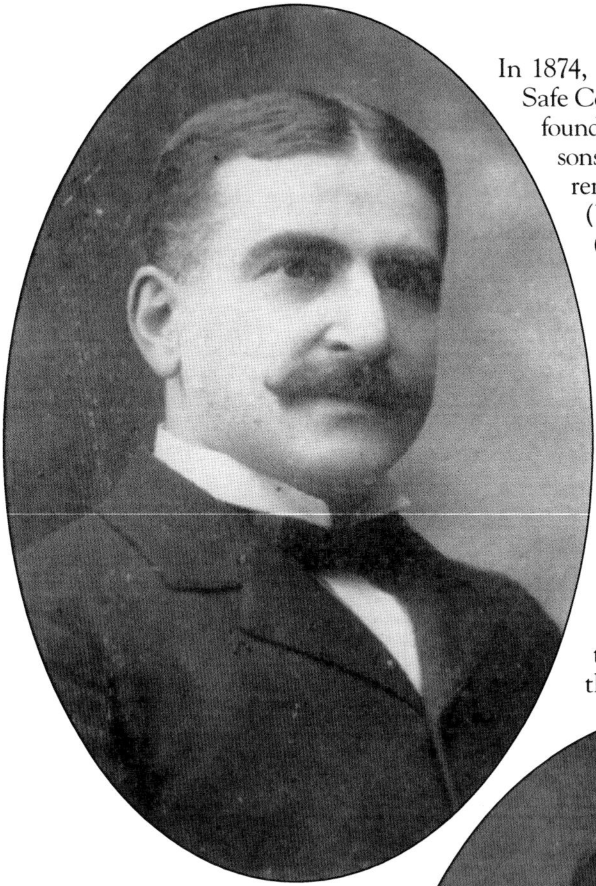

In 1874, seven years after the Mosler-Bahmann Safe Company began operations in Cincinnati, founder Gustav Mosler died at the age of 58. His sons Moses and William took over the firm and renamed it the Mosler Safe Company. Moses (below) became president, and William (left) served as vice president and treasurer. The business thrived; by the 1880s, it had sales offices in several American cities and in Mexico. By the time the Mosler Safe Company moved to Hamilton in 1891, it was an international success. Moses was a mechanical genius and inventor of the company's screw door safe. William maintained his residence in Cincinnati and made daily trips to Hamilton to manage the business. The two brothers worked to provide employees with affordable housing near their plant, created a mutual aid society to assist ill staff, and sought to improve their workers' quality of life.

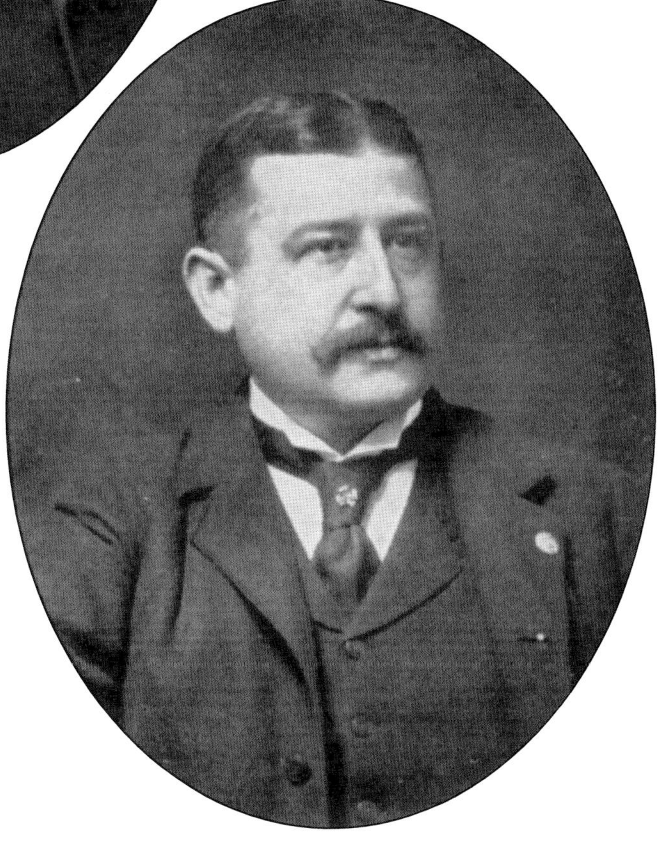

A group of civic leaders headed by Lazard Kahn and Oakey Parrish lured the Mosler Safe Company to Hamilton by offering $85,000 and land for a factory and 240 residential lots for development in "East Hamilton." Mosler's plant, built in 1891, was expanded to become the largest safe-manufacturing factory in the world. It had its own electric power plant and fire department. The company offices and landmark clock tower were built in 1893.

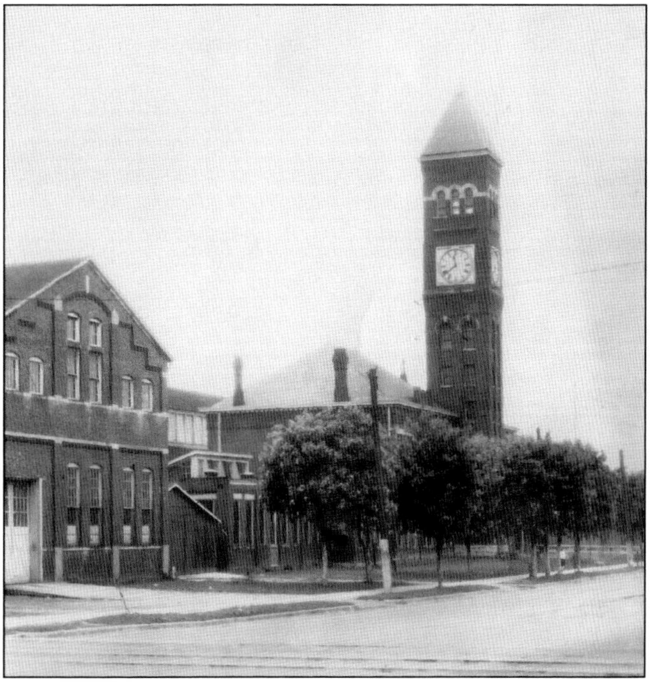

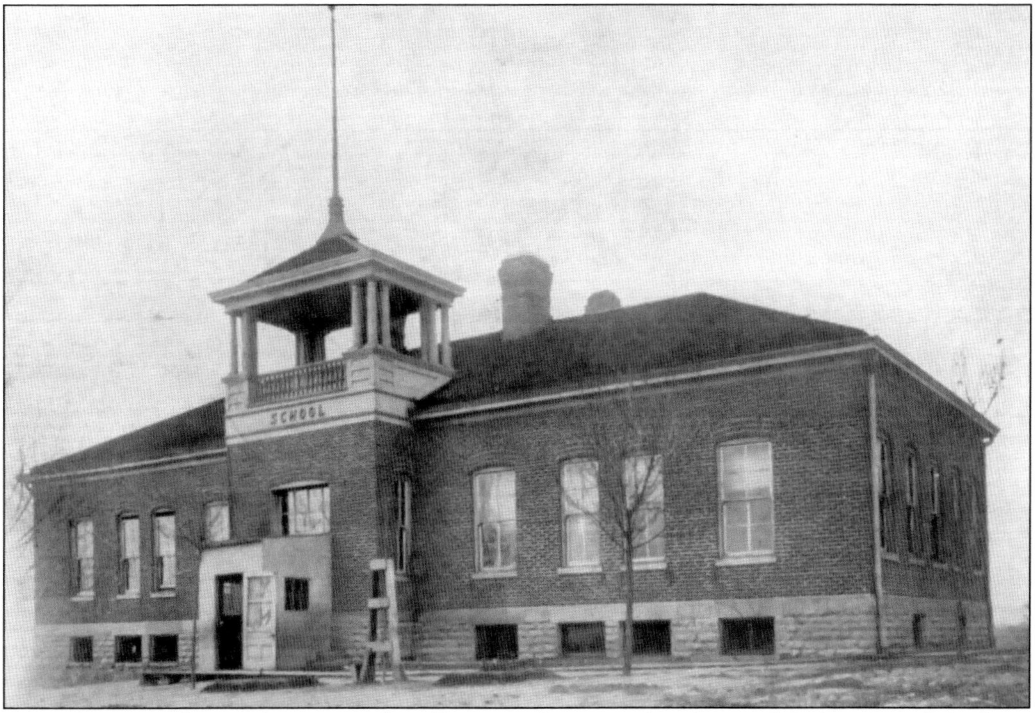

A syndicate made up of Moses and William Mosler, Lazard Kahn, and Oakey Parrish laid out lots and streets and built houses for employees. The company subsidized the building of two churches and a social hall. Hamilton's school board built the two-room school shown here at a cost of $4,927. The school was placed into service in 1900 and was renamed Van Buren School in 1909. (Courtesy of Lane Public Library.)

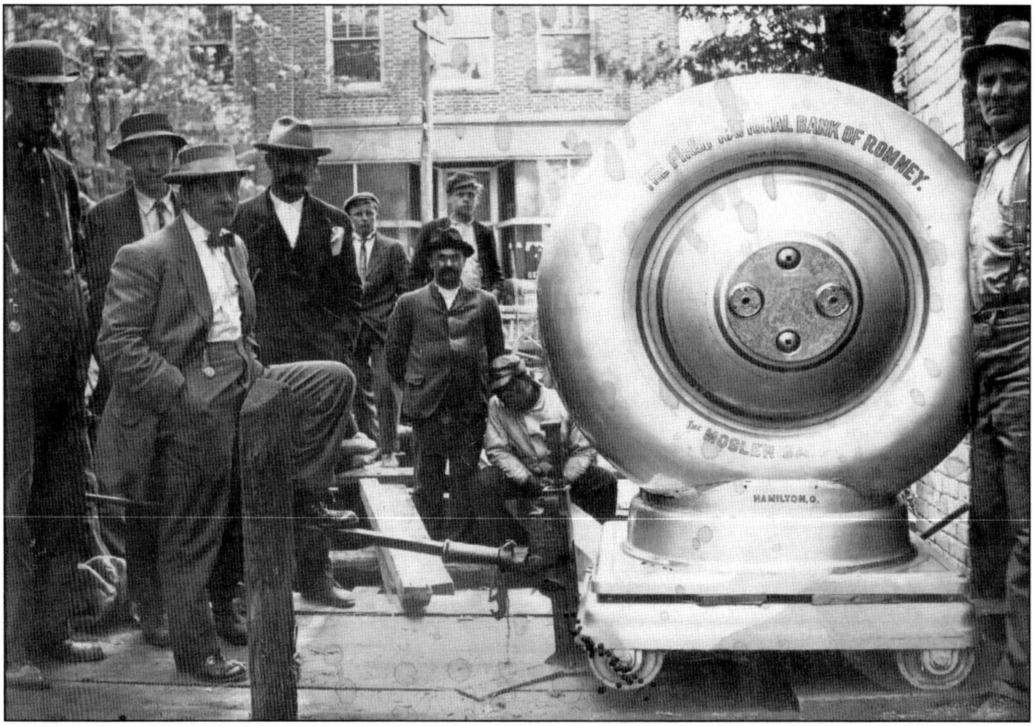

Mosler acquired the Corliss Safe Company of Providence, Rhode Island, in 1896 and moved all of its equipment to Hamilton to continue the brand as Mosler-Corliss Bank Safe until about 1910. The models sold by Mosler ranged in size from 32 to 64 inches and weighed between 8,000 and 30,000 pounds. The Mosler-Corliss Spherical Bank Safe, shown here about 1910, was manufactured for the First National Bank of Romney, West Virginia.

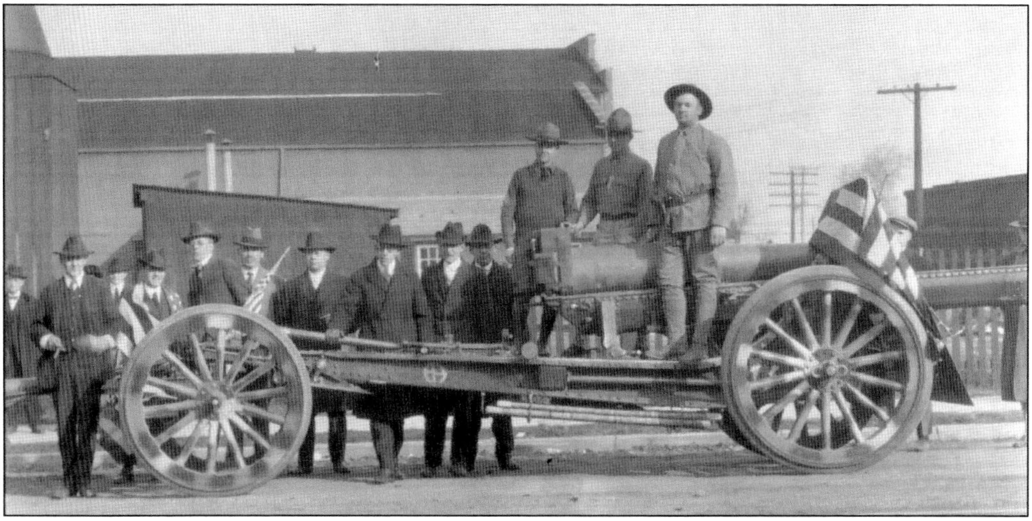

In the fall of 1917, the War Department selected Mosler to manufacture 55-millimeter Howitzer gun carriages for use during World War I. Rather than halt production of safes for customers, the company decided to erect two new buildings. A series of delays put off the completion of the first gun carriage (pictured) until September 1918. Before production of the gun carriages could start, however, an armistice was declared and the war ended.

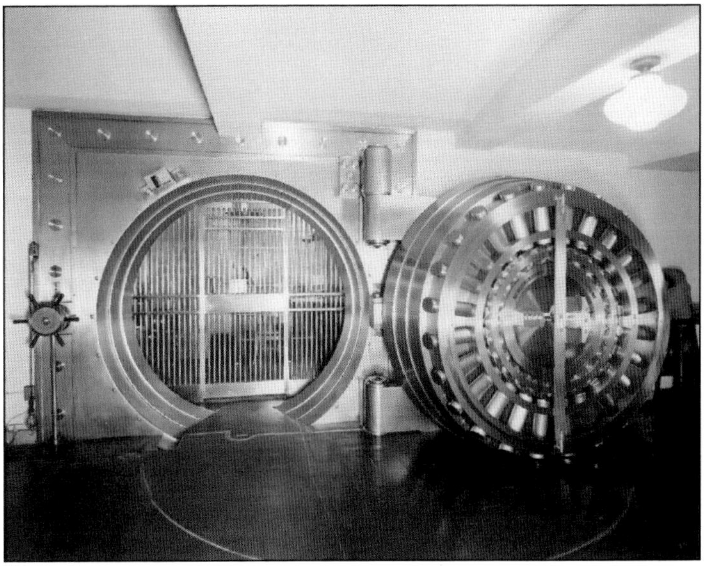

Mosler used special metal alloys with exceptional resistance to drills and torches in the construction of this large circular door and safe-deposit vault for the Federal Reserve Bank of Atlanta in the late 1920s. During the decade, the company manufactured vaults and additional security equipment for nine other Federal Reserve Banks or branches in Birmingham, Nashville, Denver, St. Louis, New Orleans, Detroit, Oklahoma City, Baltimore, and San Francisco.

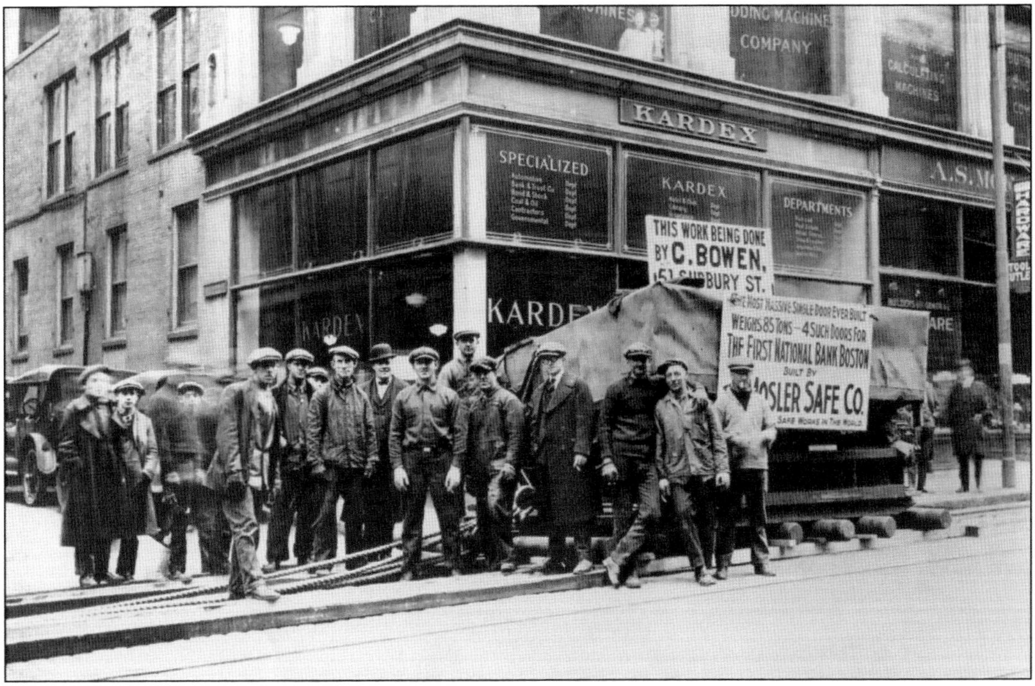

These men are hauling a section of an 85-ton vault door over wooden rollers to the First National Bank of Boston on November 15, 1923. The trip to transport four vault doors from the Albany Street railroad yard to the bank on Federal Street took more than 12 hours. The sign at right claims the men are moving "the most massive single door ever built."

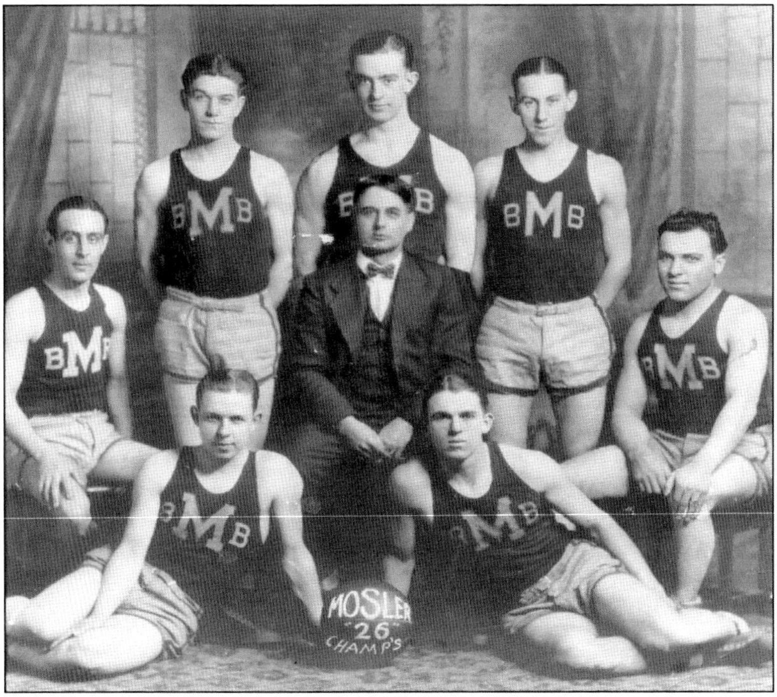

For many years, Mosler supported the recreational activities of its employees. Members of the company's state industrial champion basketball team of 1927 pose for a photograph. The team's players were, from left to right, (first row) Bud Kinkey and Clarence Shafer; (second row) Carlisle Motzer, Charles Williamson (coach), and Ben Blumenthal; (third row) Shine Morton, Ralph Stephenson, and Ike Phipps.

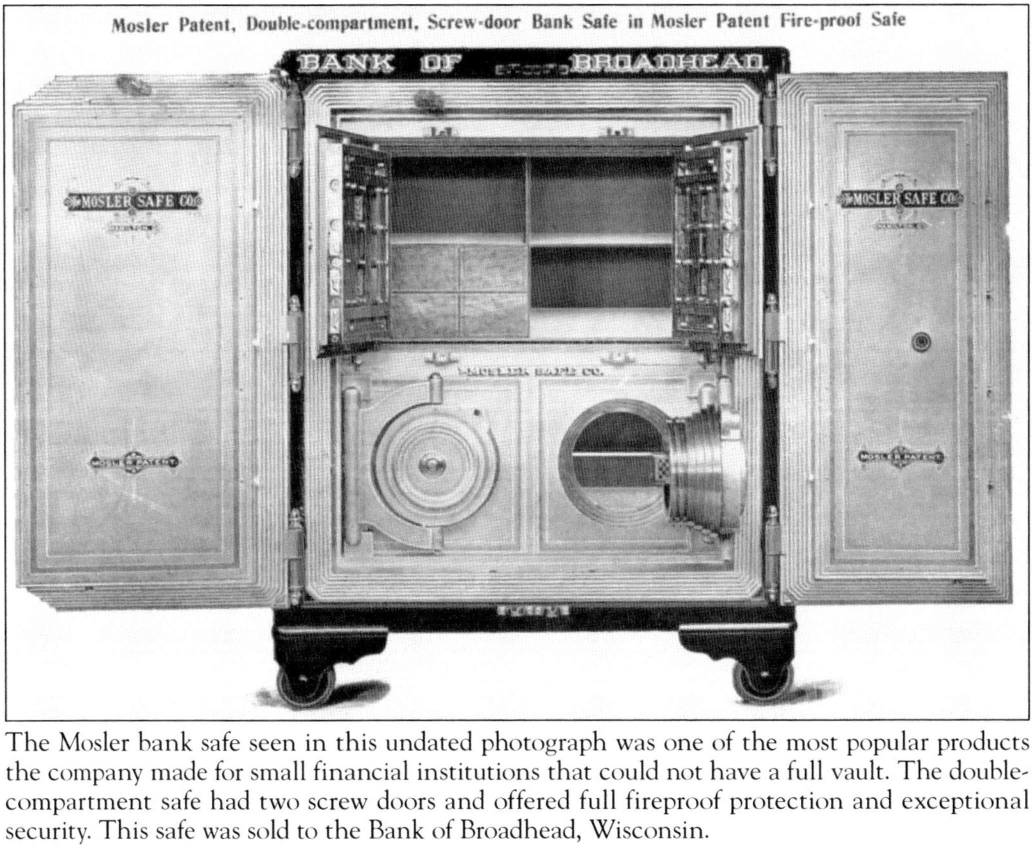

The Mosler bank safe seen in this undated photograph was one of the most popular products the company made for small financial institutions that could not have a full vault. The double-compartment safe had two screw doors and offered full fireproof protection and exceptional security. This safe was sold to the Bank of Broadhead, Wisconsin.

During the depression of the 1930s, Mosler solicited subcontract work from other metalworking companies in southern Ohio in order to give their employees a few days work each week. The R.K. LeBlond Machine Company in Cincinnati contracted with Mosler for the manufacture, machining, and assembly of 500 lathe aprons. These men are working to assemble parts of the lathes into finished products.

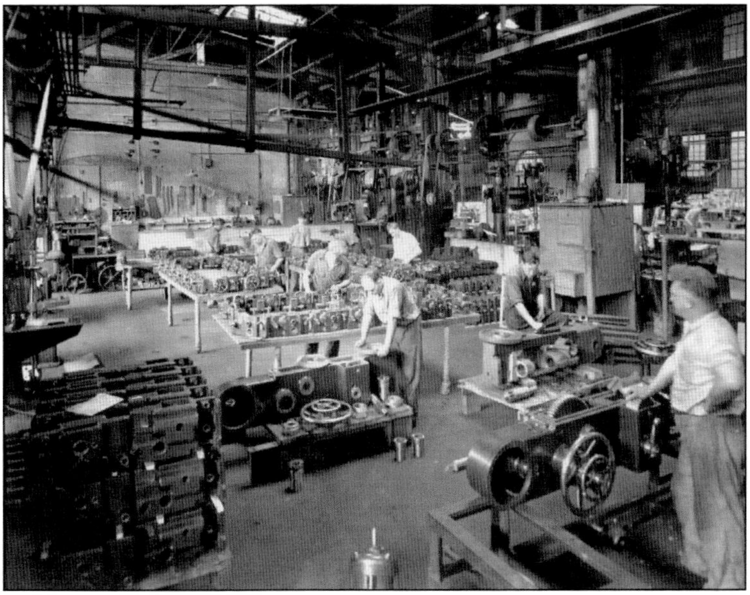

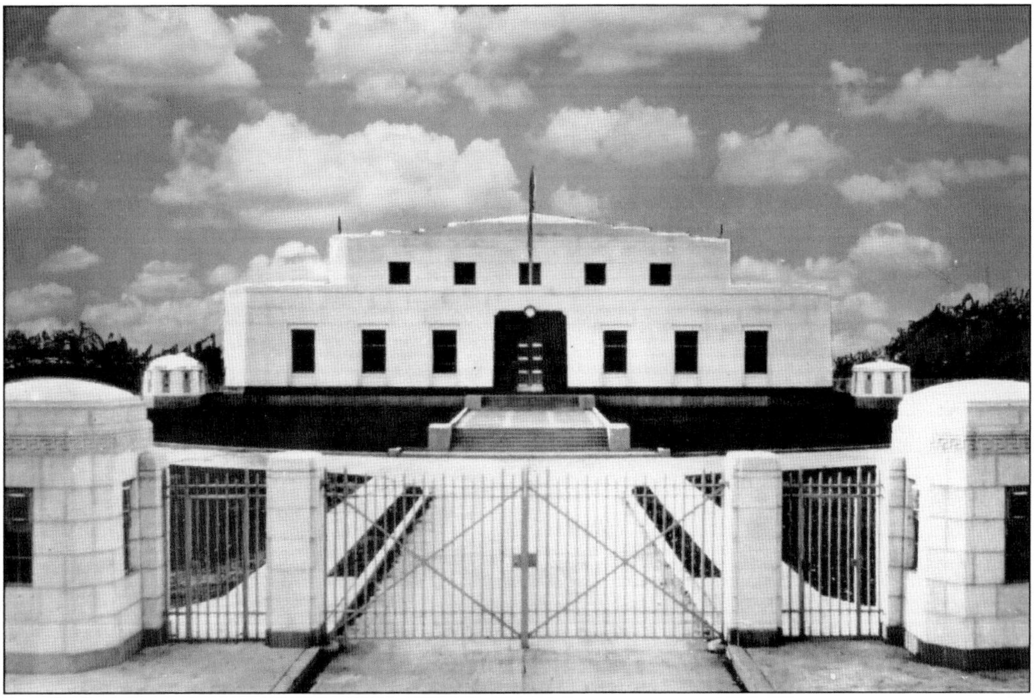

In 1935, the federal government ordered sets of 21-inch thick, 20-ton vault doors for the newly constructed gold depository (pictured) at Fort Knox, Kentucky. After the depository opened in 1937, as much as $20 trillion worth of gold—about half of the world's supply at the time—was safely and securely stored in vaults designed and manufactured by Hamilton's Mosler Safe works.

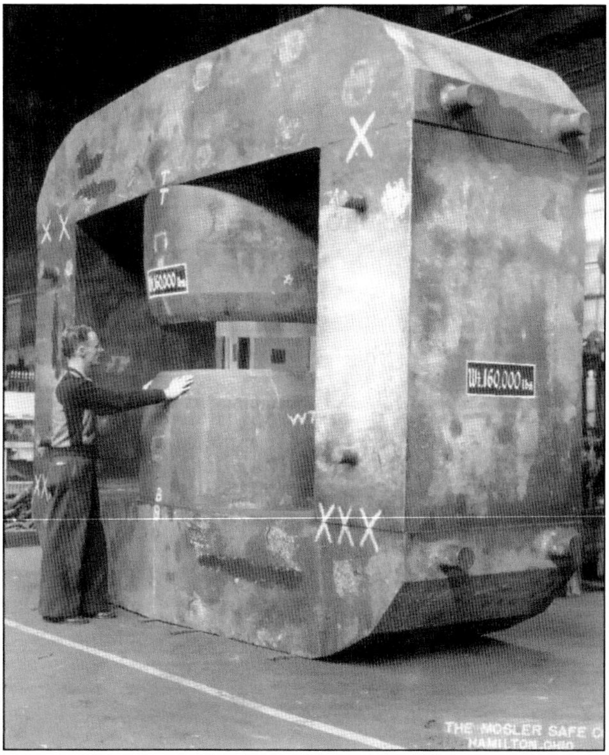

Al King examines the 160,000-pound cyclotron magnet, or "atom smasher," that Mosler made for Ohio State University in 1938. Mosler did the machine work for the magnet, and the American Rolling Mill Company in Middletown manufactured the steel casings. Several cyclotrons were made in the late 1930s for universities, research institutes, and government contractors. The cyclotrons played an important role in the research, which contributed to the development of nuclear weapons.

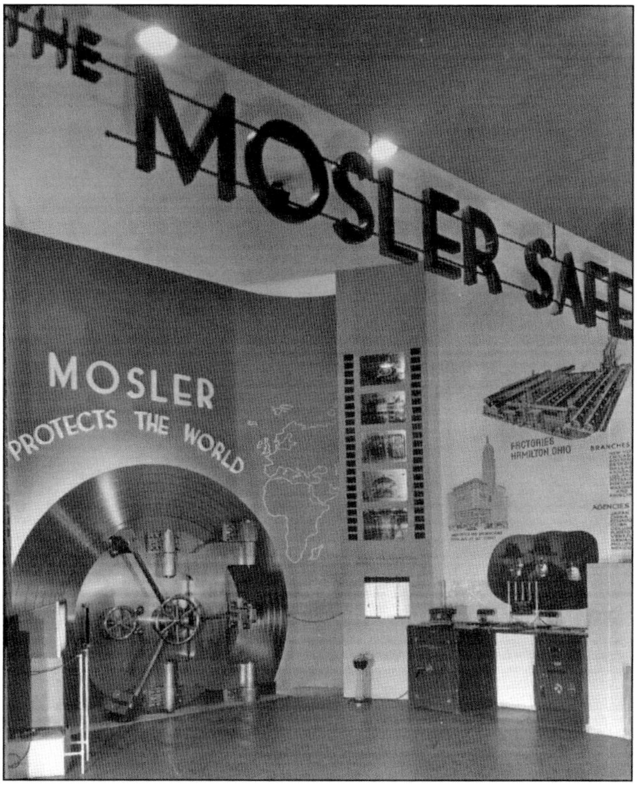

This is a portion of the Mosler Safe Company's impressive exhibit at the 1939 World's Fair in New York City. The display included the latest security equipment and an actual round vault door. One of the most unusual items displayed was a tiny steel safe the company had made to protect George Washington's false teeth for the 1932 World's Fair in Chicago.

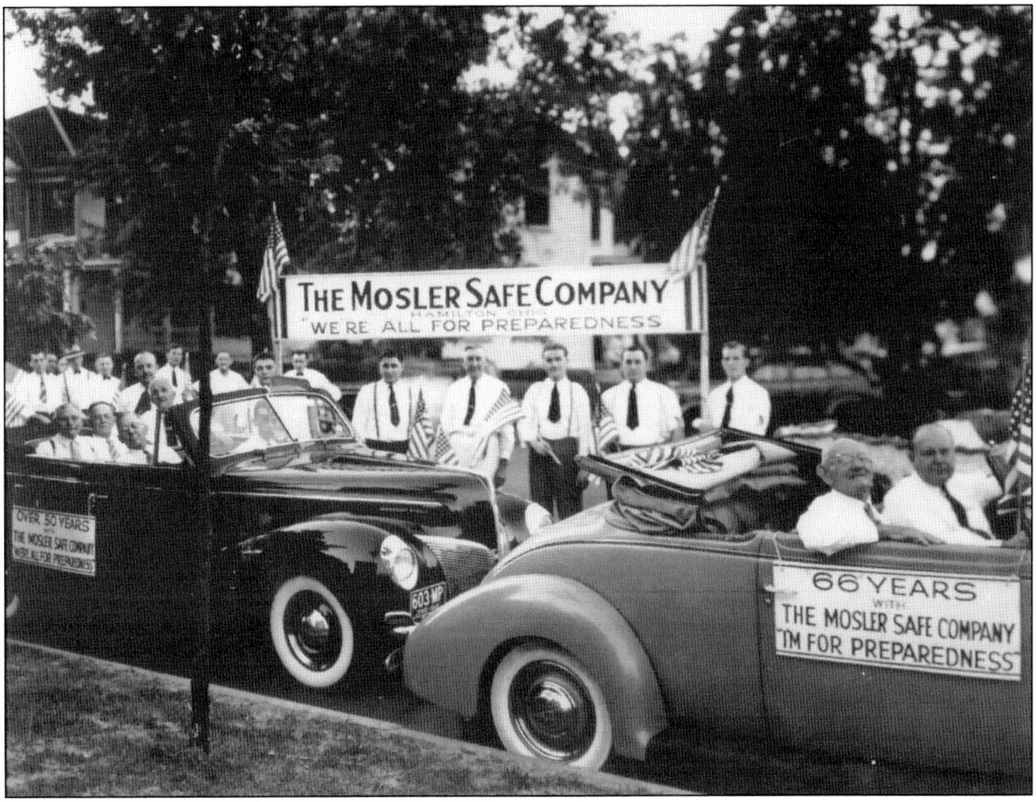

Mosler executives and employees took part in the Hamilton Preparedness Day parade on August 2, 1940. The three-day program featured patriotic displays; speeches by politicians, military leaders, and local officials; a Miss Hamilton beauty contest; and a parade that took nearly two hours to pass the reviewing stand. Mosler employees who had been with the company for more than 50 years rode in open cars down the four-mile parade route.

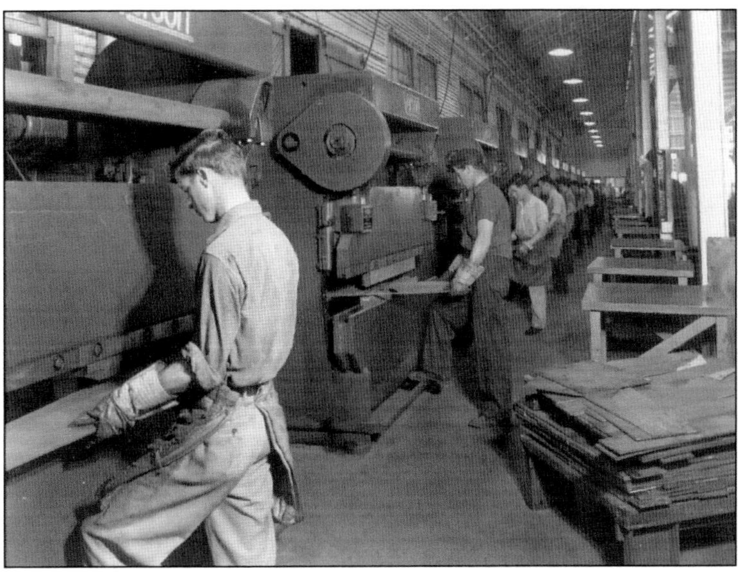

At the start of World War II, Mosler decided to erect a new building for fabricating and heat-treating armor plate needed for 37-millimeter gun shields, half-track trucks, and airplanes. Heat-treating furnaces were installed in the new structure, and the plant went into full operation before the roof was completed. In this photograph, company workers are using presses to machine the armor plates.

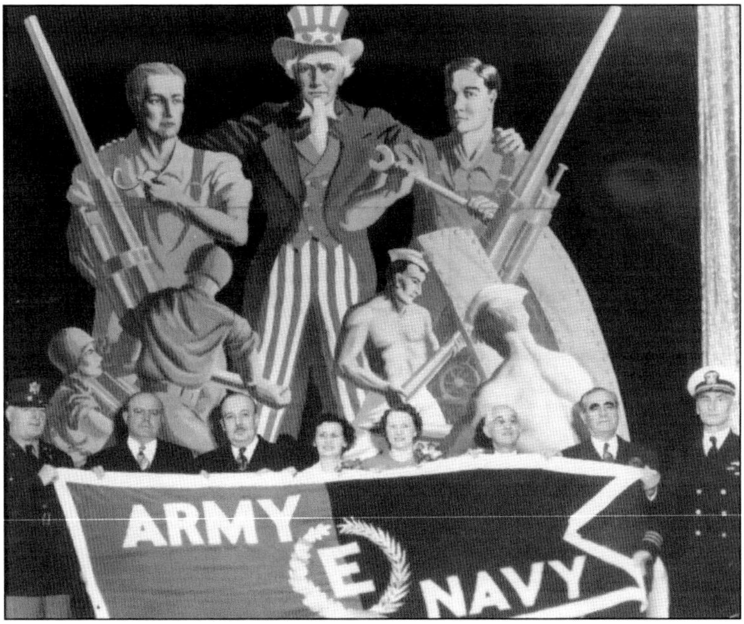

During World War II, Mosler was converted to the manufacture of wartime materials. The company received an Army-Navy "E" Award for "excellence in production" on April 17, 1943. Participants in the ceremony were, from left to right, Fred A. McMahon, Harry H. Lynn, Carl P. Bartels, Lillian Renneker, Leona Niehaus, Adolph E. Trapp, Edwin H. Mosler Sr., and W.J. Butler. Mosler earned three additional stars for its "E" award before the end of the war.

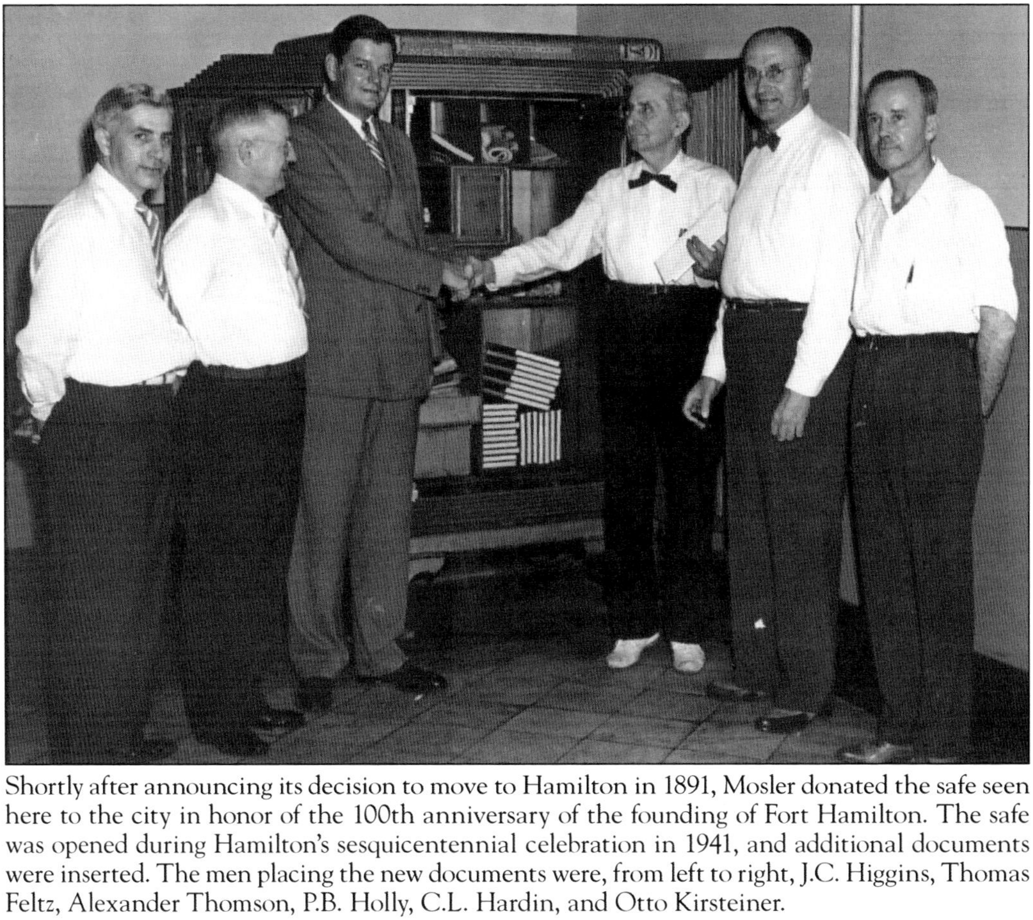

Shortly after announcing its decision to move to Hamilton in 1891, Mosler donated the safe seen here to the city in honor of the 100th anniversary of the founding of Fort Hamilton. The safe was opened during Hamilton's sesquicentennial celebration in 1941, and additional documents were inserted. The men placing the new documents were, from left to right, J.C. Higgins, Thomas Feltz, Alexander Thomson, P.B. Holly, C.L. Hardin, and Otto Kirsteiner.

Here, two unidentified men stand next to the fifteen-foot-tall, five-foot-thick doors Mosler designed and built for the Atomic Energy Commission Oak Ridge National Laboratory in 1959. The company had earlier built the protection system for America's "Charters of Freedom"—the Declaration of Independence, the Constitution, and the Bill of Rights—and the vault for storing them at the new National Archives building, which opened in 1952.

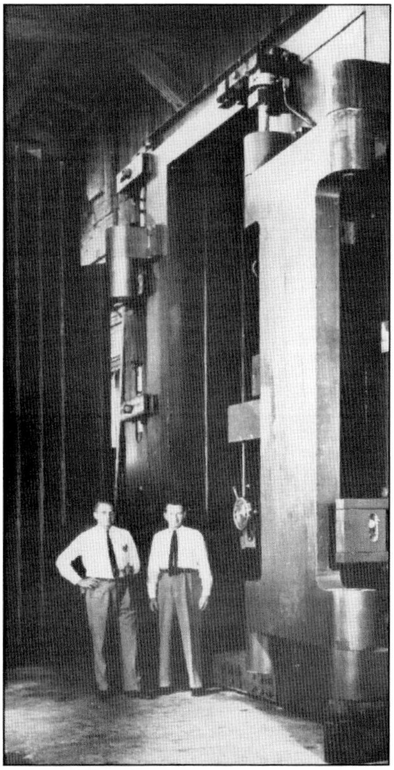

The Mosler-Duplex Auto-Teller, or "Snorkel," is seen in operation on a street in Newark, New Jersey, around 1952. The company developed the curbside teller to help banks such as Fidelity Union Trust meet the needs of their customers who preferred to drive rather than walk to their bank. The device combined circuit television with a thief-proof currency tray, two-way speaker, and bandit alarm. Duplex Electric Company was a Mosler associate.

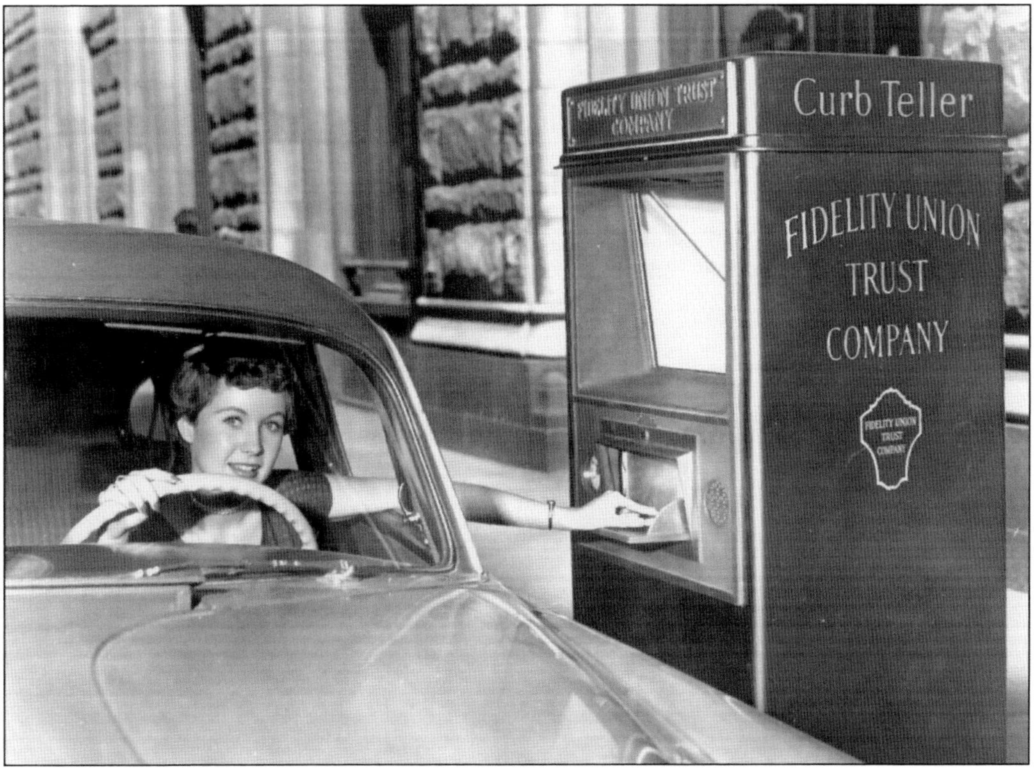

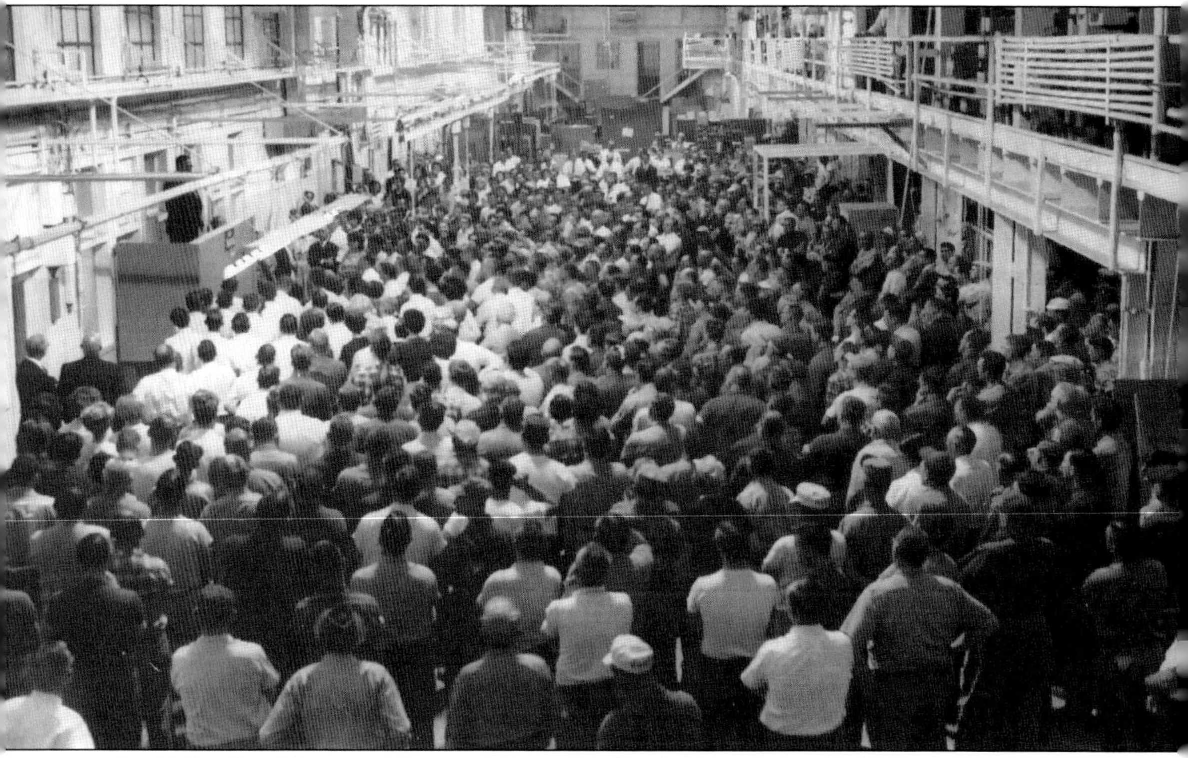

In 1962, after 95 years of being 100-percent family owned, Mosler began selling stock to the public in order to acquire capital to compete and grow. Within five years, chairman of the board John Mosler, a great-grandson of the company's founder, and president Bill Marquard informed the assembled employees seen in this May 1967 photograph that Mosler had been sold to the American-Standard Company. Production continued at the plant. In 1986, a group of company managers and a private New York investment bank purchased the company. The Grand Avenue plant was closed in 1996, and safe manufacturing in Hamilton ended. The company kept its headquarters in town and operated nearly 60 service centers around the country. By August 2001, however, the company had such a massive debt burden that it was forced to stop all operations, liquidate its assets, and declare bankruptcy.

Four

Engines and Vehicles

The initial thrust into manufacturing engines in Hamilton came in the 1850s, when Job Owens, Jacob Ebert, Elbridge Dyer, and Clark Lane were first operating their businesses. The first three men had founded an iron-stove factory in 1845, and Lane began blacksmithing in Hamilton in 1844. After Ebert died in 1854, Lane bought into the firm, and Owens, Lane and Dyer began manufacturing threshing machines, papermaking equipment, sawmills, hay rakes, and other farm machinery at the firm's Eclipse Machine Works. In 1858, the company manufactured the first practical wood steam engine and threshers. During the 1870s, Owens, Lane and Dyer was building self-propelled steam-traction engines used to plough fields, pull loads over roads, and provide portable power at a particular location.

In 1876, John C. Hooven was selling farm implements and his Monarch line of portable traction engines and threshers that he had designed and built at a storefront in downtown Hamilton. When Owens, Lane and Dyer went into receivership, Hooven formed a partnership that bought the company in 1880. The newly formed company was named Hooven, Owens, Rentschler. In 1883, the company manufactured its first Corliss stationary steam engine and soon specialized in making Corliss engines exclusively. The company's Corliss Engine Works became one of the largest engine makers in the country. In 1928, it was merged with the Niles Tool Works to create the General Machinery Corporation. Other mergers joined the company with two railroad locomotive manufacturers in the 1940s and 1950s.

The Ritchie and Dyer Company was formed in 1881 to manufacture traction engines, smaller tram engines, and sawmills. William Ritchie organized the Advance Manufacturing Company in 1887 to sell gasoline internal combustion engines of his design; by 1899, he had built the first gas-powered vehicles in Hamilton.

Other early vehicle makers, including Columbia Carriage Company, built buggies, surreys, and phaetons. Several Hamilton companies made automobiles, including the Republic Car Company, which had ties with the Rentschler family. More recently, Ford Motors operated a plant in Hamilton for more than 30 years.

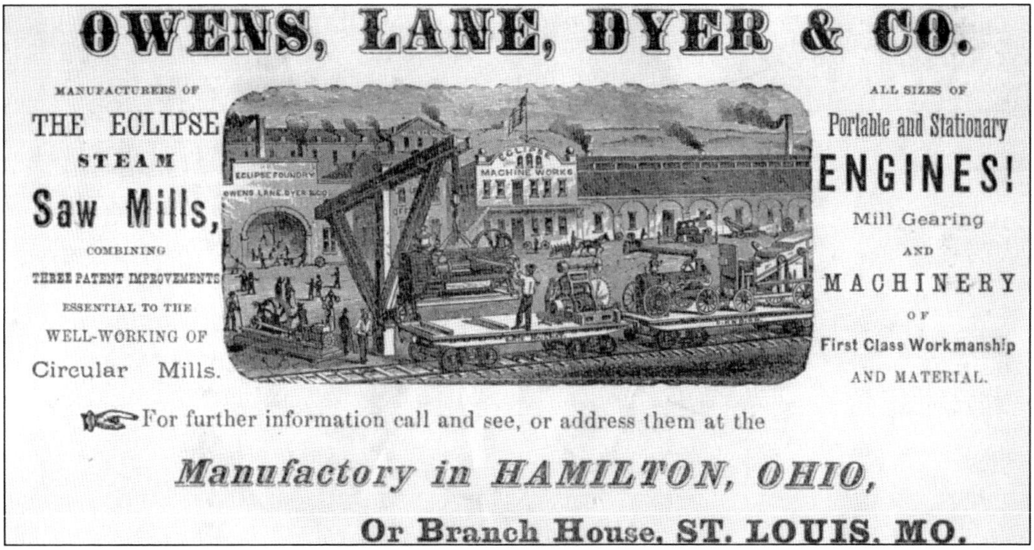

Job Owens, Jacob Ebert, and Elbridge Dyer founded their factory as the Hamilton Agricultural Works on the hydraulic in 1845. After Clark Lane bought Ebert's interest in 1854, the firm became the Eclipse Machine Works of the Owens, Lane and Dyer Company. This 1870 advertisement features an artist's rendering of the company's plant, where award-winning sawmills, steam engines, and threshing machines were made.

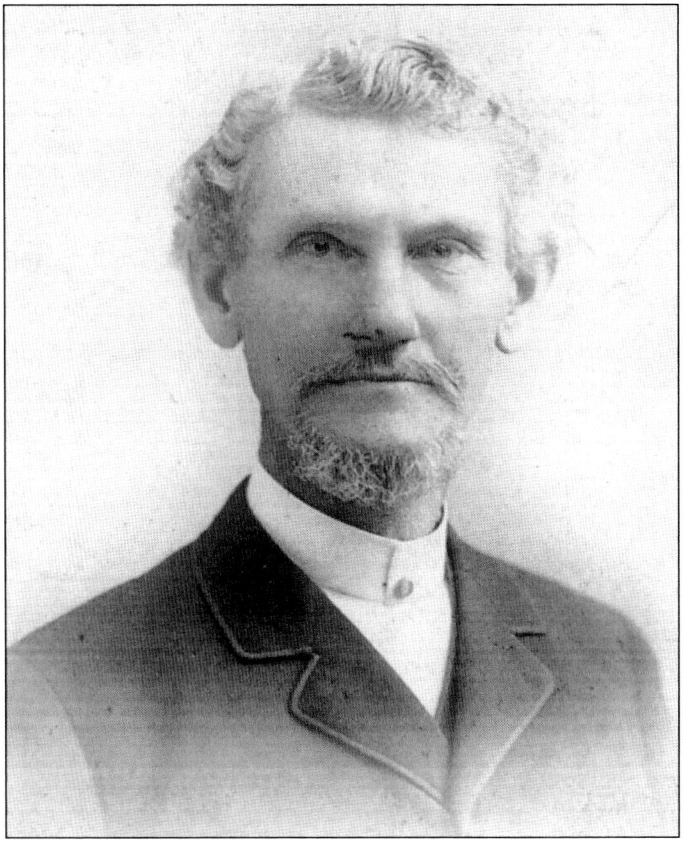

Clark Lane had a very profitable blacksmith business when he joined Owens and Dyer in 1854. During the 1860s, he built two octagonal buildings on Third Street, one for a home, and the other, which he stocked with more than 2,000 books, he gave to the city for a public library. Lane also bought the farm and home owned by Elbridge Dyer and gave it to Butler County for use as an orphanage and children's home.

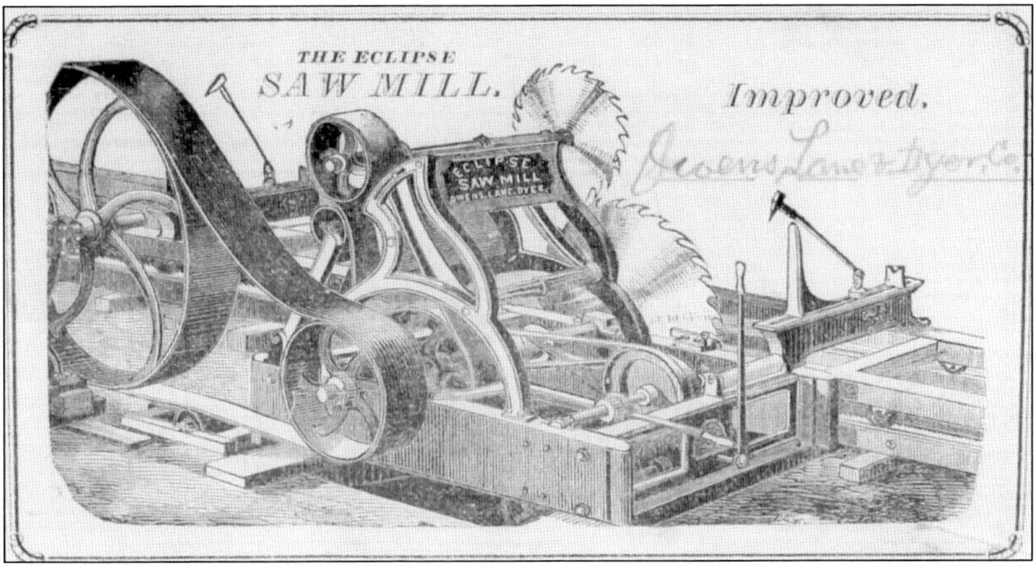

These two images are taken from pages in an 1870 catalog for products manufactured by Owens, Lane and Dyer. Shown above is the company's Eclipse sawmill mounted on a skid. The machine shown below is the Hamilton thresher and separator, marketed as the best grain thresher in the world. Despite the success of the company, it was in financial trouble by 1874 due to uncollected accounts and a general economic slump with record numbers of farm foreclosures. Elbridge Dyer committed suicide, and in 1876, the company went into receivership, with Clark Lane appointed the receiver. In 1879, Owens, Lane and Dyer was out of business entirely. The defunct company's plant on the hydraulic was deeded to Job Owens, who had gone into partnership with J.C. Hooven, George Rentschler, and others. Lane, acting as Owens, Lane and Dyer's receiver, also deeded a factory location to William Ritchie and William Dyer, who started a company in a plant at Fourth and Vine Streets.

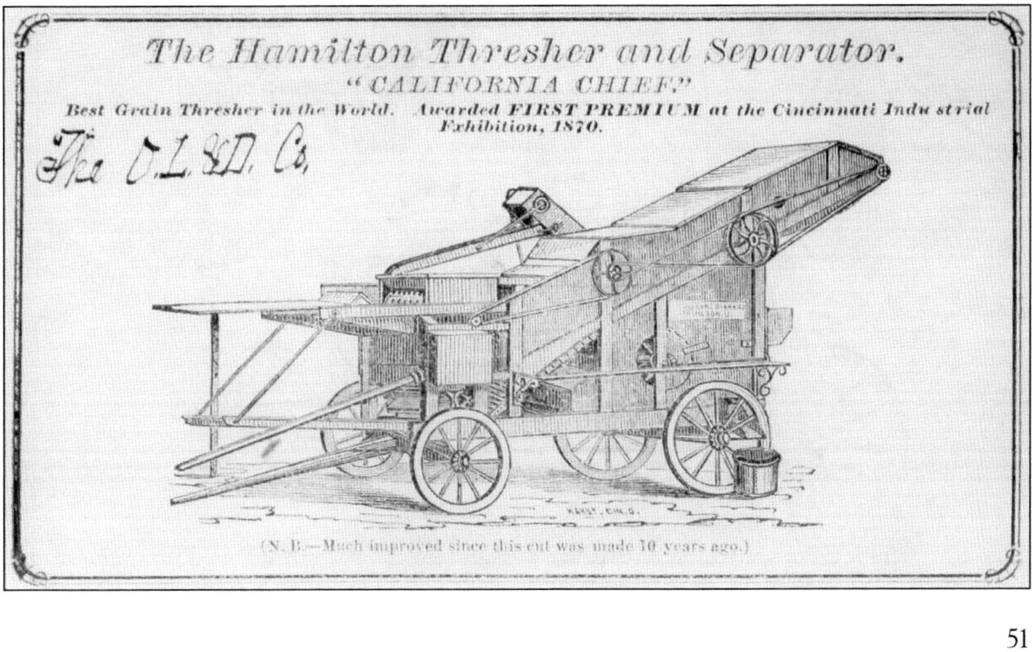

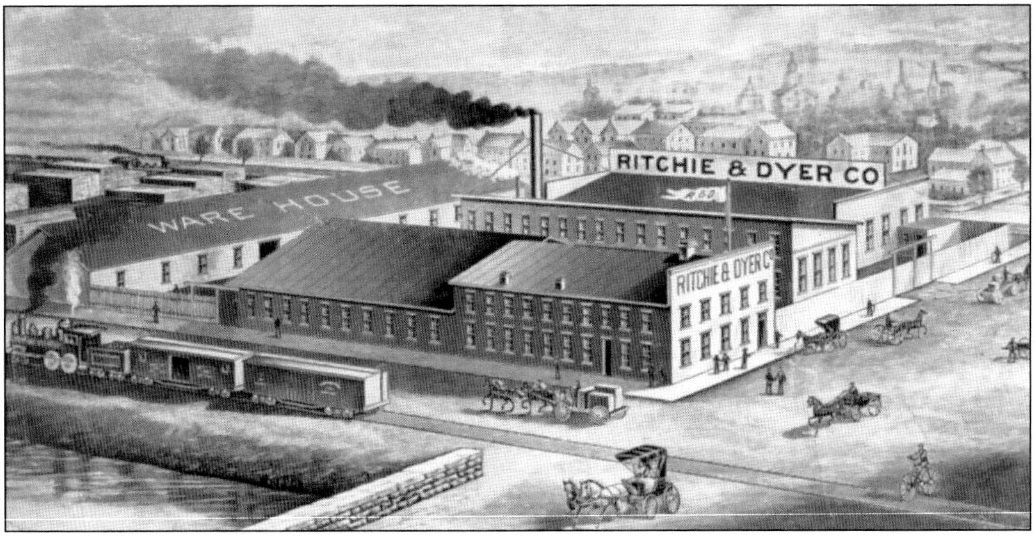

In 1881, William Ritchie and William Dyer located their factory, seen in this fanciful rendering, on Vine Street along the hydraulic between Fourth and Lowell Streets. The company made self-propelled steam-traction engines, sawmills, ice-cutting equipment, and small machinery. Dyer, a son of Elbridge Dyer, one of the owners of Owens, Lane and Dyer, departed the company in 1882, leaving Ritchie in full control.

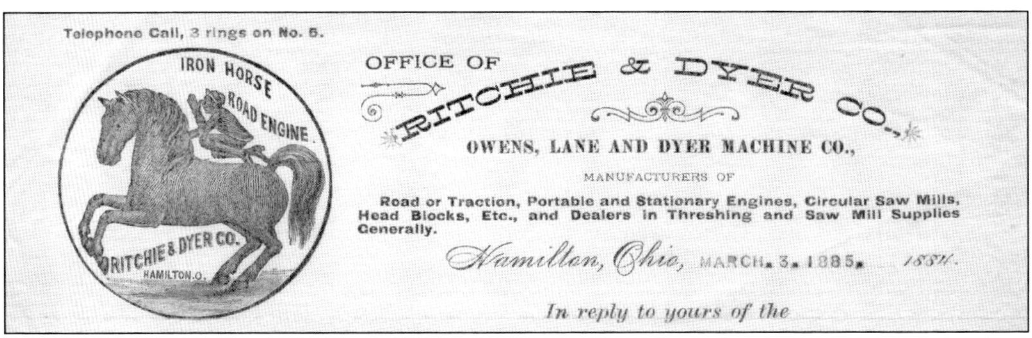
Telephone Call, 3 rings on No. 5.

IRON HORSE ROAD ENGINE

RITCHIE & DYER CO.
HAMILTON, O.

OFFICE OF
RITCHIE & DYER CO.,
OWENS, LANE AND DYER MACHINE CO.,
MANUFACTURERS OF
Road or Traction, Portable and Stationary Engines, Circular Saw Mills, Head Blocks, Etc., and Dealers in Threshing and Saw Mill Supplies Generally.

Hamilton, Ohio, MARCH. 3. 1885.

In reply to yours of the

Ritchie and Dyer highlighted the firm's ties with the defunct Owens, Lane and Dyer Company in the firm's letterhead (pictured). By 1885, Hooven, Owens & Rentschler was specializing in making only Corliss engines and had relinquished its rights to repair Owens, Lane and Dyer's engines. Ritchie and Dyer took over the repair of the older company's Eclipse engines and manufactured their own engines, threshers, and sawmills.

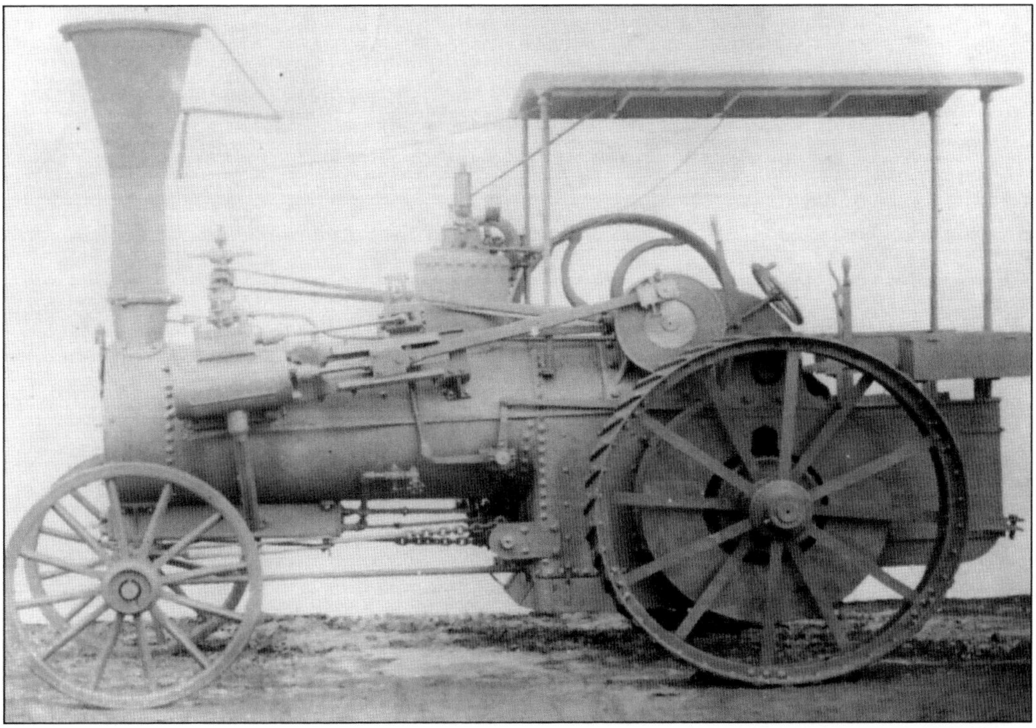

Ritchie and Dyer manufactured this steam road traction engine in 1885. While many of the early traction engines were horse-steered, this road engine had a driver's seat and steering wheel. The engine was almost identical to the tractors built by Owens, Lane and Dyer. This photograph was quite likely used for advertising the company's road engines.

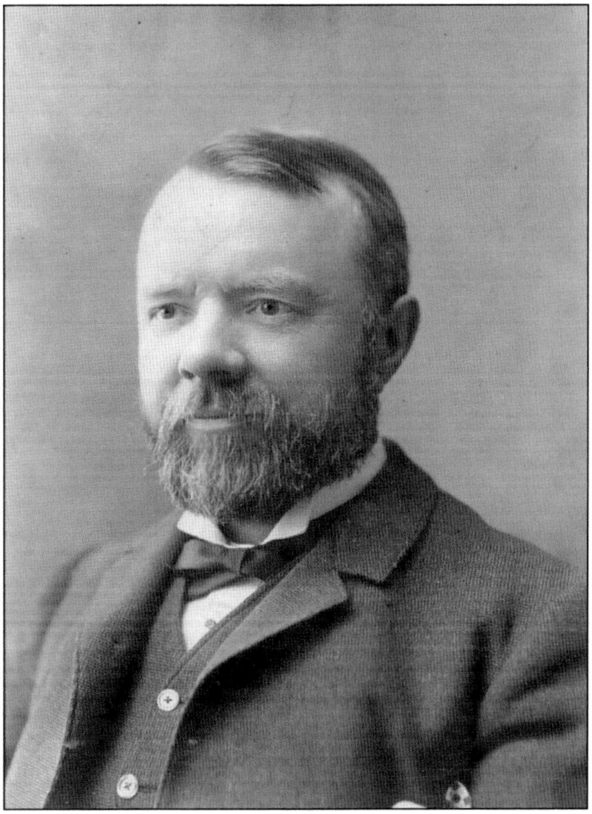

William Ritchie formed the Advance Manufacturing Company in 1887 to build the gasoline internal combustion engines he designed while still conducting Ritchie and Dyer's business. In 1892, Pres. Benjamin Harrison appointed him commissioner of the World Columbian Exposition in Chicago. When Ritchie returned to Hamilton, he closed down Ritchie and Dyer Company to concentrate his efforts on his gas engine company. He died in 1906, and his son Oscar continued the business.

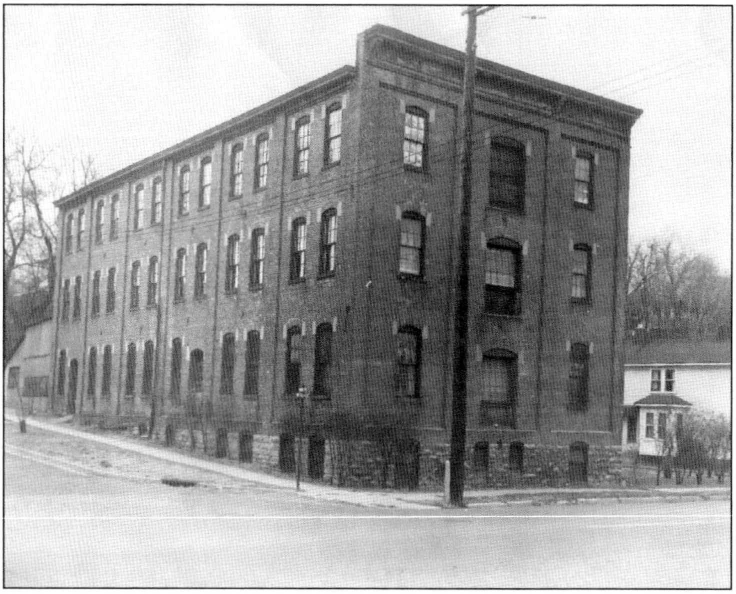

The Advance Manufacturing Company works were housed in this building at North B and Wayne Streets. The firm made a variety of products here, including gasoline engines for farm machinery and road pavers. In 1899, the firm built the Motor Wagon, the first gas-powered car in Hamilton.

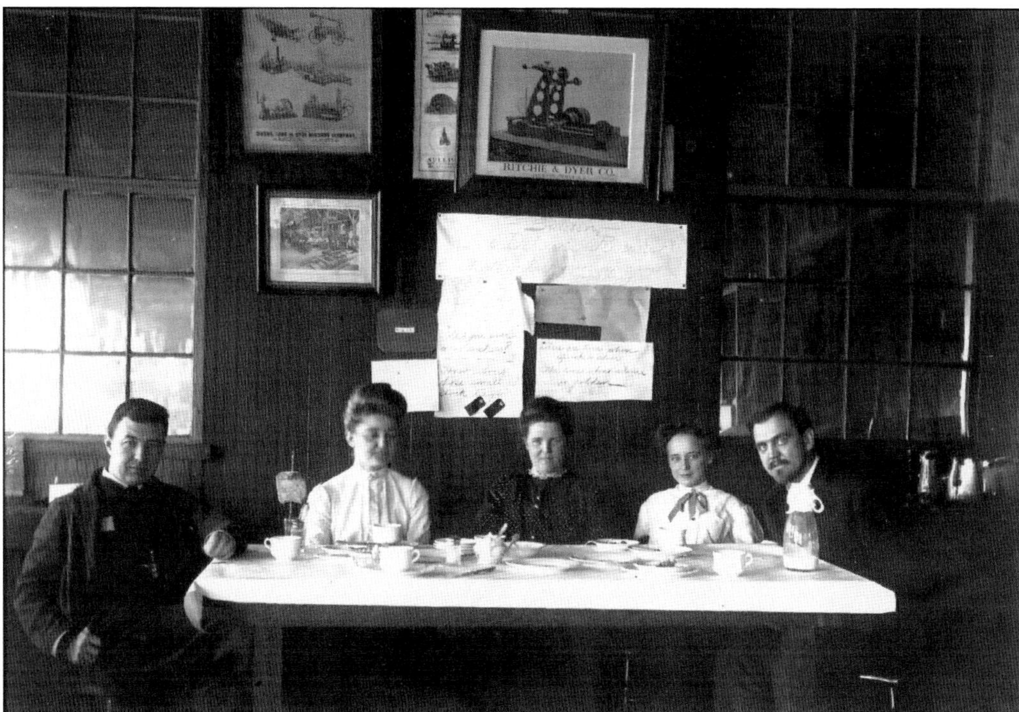

Advance Manufacturing Company claimed it had the first factory lunchroom. Seated around the table in this 1903 photograph are, from left to right, Oscar N. Ritchie, Emma Ritchie, Margaret Kane, Myrtle Buckardt, and Paul Buckardt. The framed pictures on the wall feature Owens, Lane and Dyer and Ritchie and Dyer advertisements. Several slogans are written on the white sheets of paper tacked to the wall, the smallest of which asks people to "Keep Tidy."

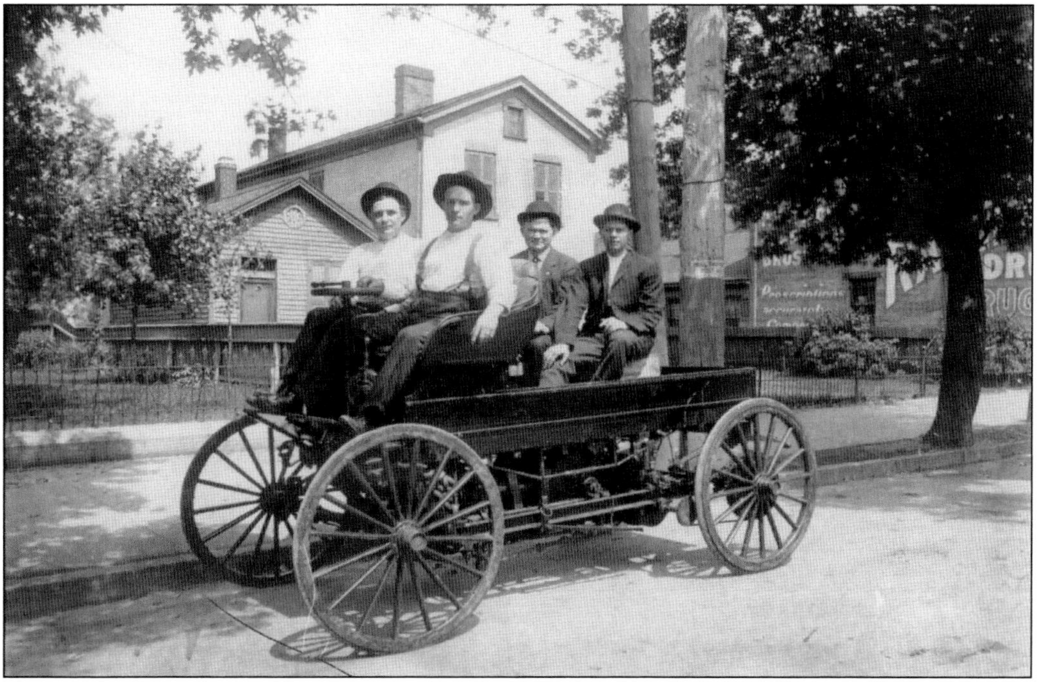

William Ritchie and his son Oscar built this truck in 1898, making it one of the earliest vehicles manufactured in Hamilton. It was powered by a 10-horsepower gasoline engine they had made in the Advance Manufacturing plant. The men in the truck are, from left to right, (front seat) John Willenbrink and Jack Nicholas and (back seat) Ted Tillman and Otto Zollers.

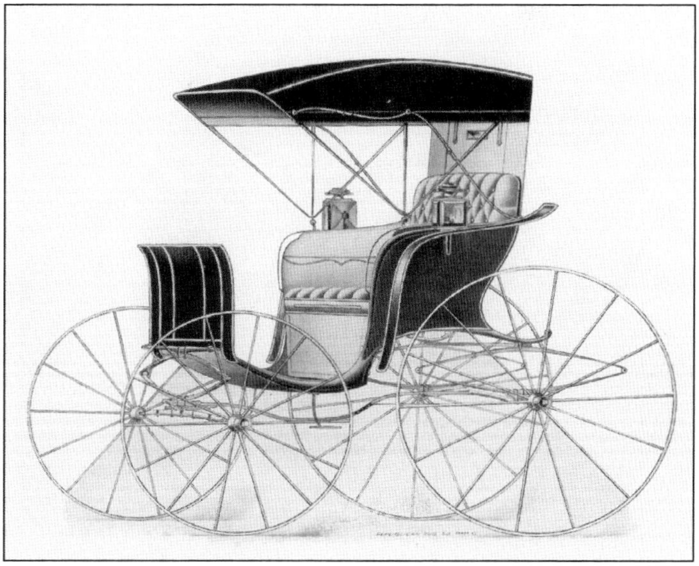

Columbia Carriage Company, formed in 1890 and incorporated in 1898, made thousands of buggies, surreys, and phaetons, like the one seen here at its factory on Central Avenue. The company offered customers high-quality horse-drawn carriages that were customized by a variety of options in seats, wheels, tops, springs, and axles. Advertisements for the company asked, "Why buy just as good, when you can get a genuine 'Columbia' for about the same money?"

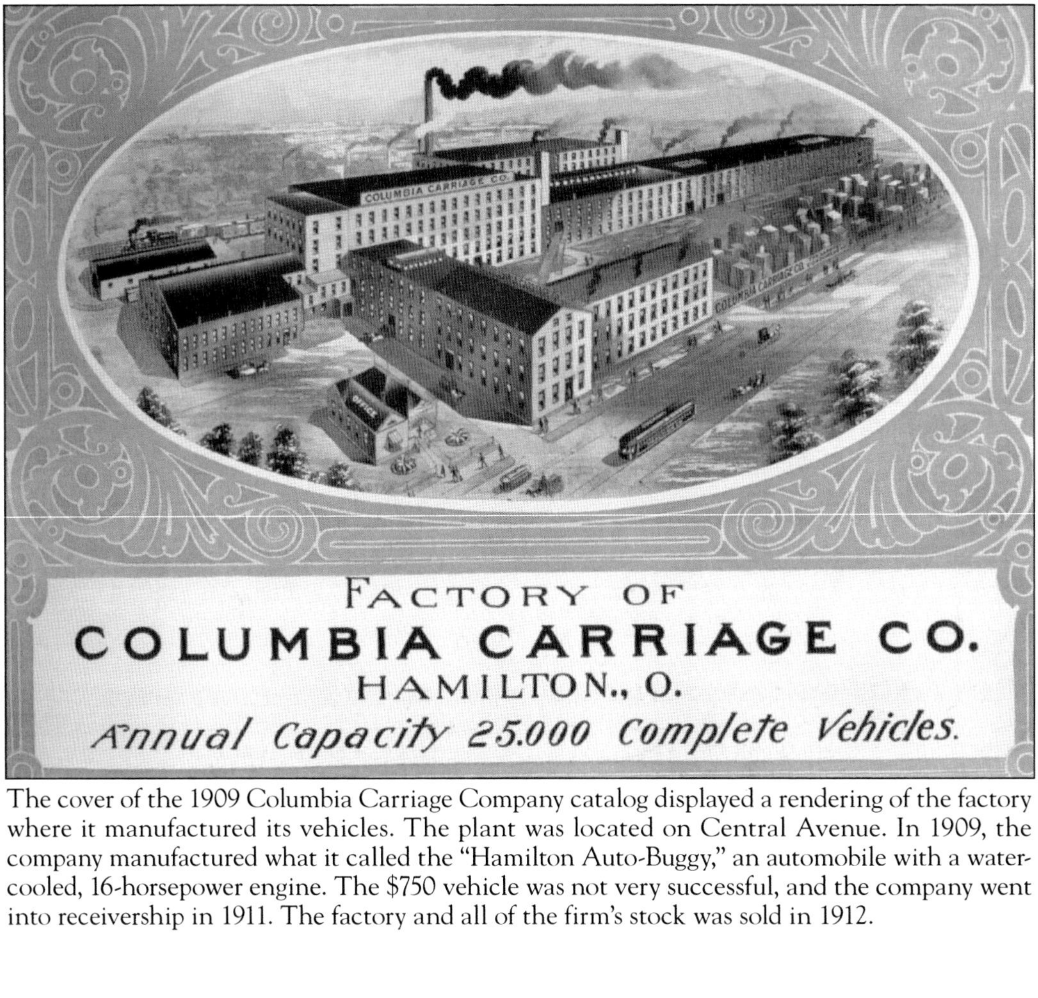

The cover of the 1909 Columbia Carriage Company catalog displayed a rendering of the factory where it manufactured its vehicles. The plant was located on Central Avenue. In 1909, the company manufactured what it called the "Hamilton Auto-Buggy," an automobile with a water-cooled, 16-horsepower engine. The $750 vehicle was not very successful, and the company went into receivership in 1911. The factory and all of the firm's stock was sold in 1912.

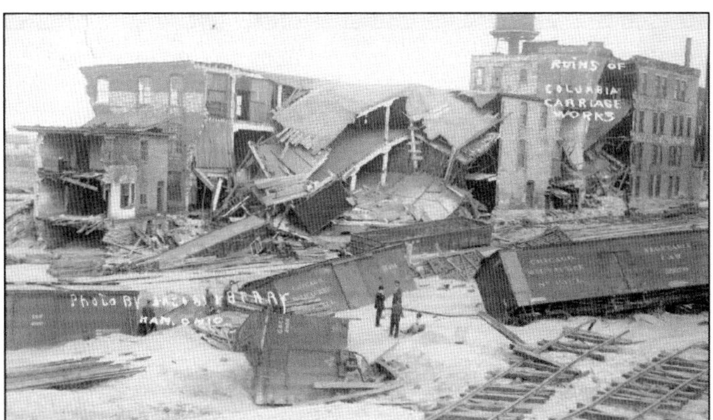

The 1913 Great Miami River flood caused massive destruction to the building and inventory of the New Columbia Carriage Company, the firm that was formed after the bankruptcy sale in 1912. The new owners were unable to recover from the flood and went out of business. The factory was rebuilt; by 1916, it was occupied by the Fischer Can Company.

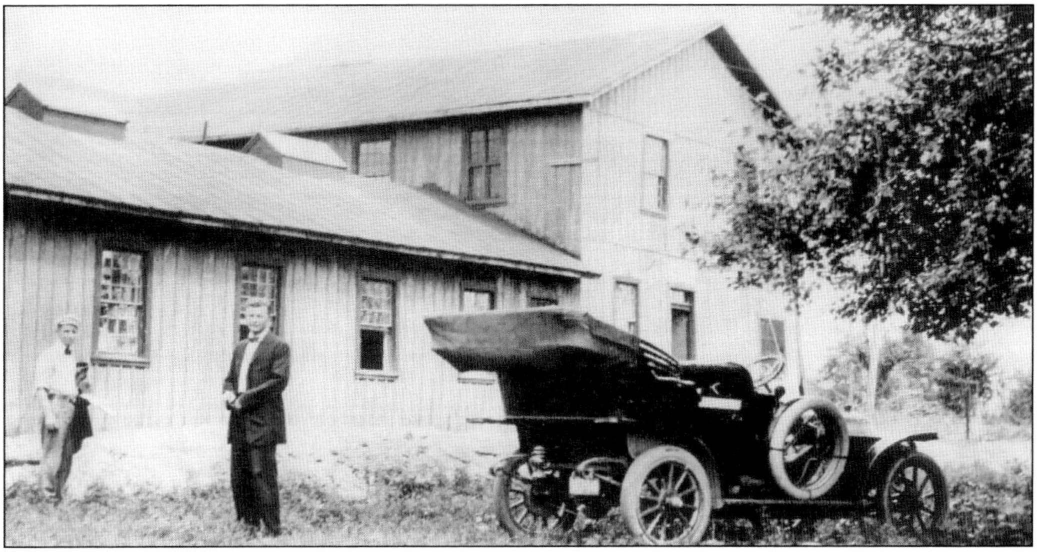

Above, Frederick Brant Rentschler (right) and George Stanley Helvey stand outside Louis Snider's old Fairgrove paper mill in 1908. The mill was converted into an automobile factory. Founders of the Republic Motor Car Company were George Adam Rentschler, Charles U. Carpenter, George H. Helvey, Clarence Knowles, Cordy Cousins, Gordon S. Rentschler, and Frederick Rentschler. The company built two cars in 1909, thirty in 1910, and One hundred fifty in 1911. By 1912, a new factory was built next to the old one, and the company produced several hundred vehicles there until 1914. Republic automobiles were very popular in Hamilton. The passengers in the back seat of the touring car in the photograph below are Burt R. Skinner (left) and Virginia Skinner. Frederick Brant Rentschler is the driver. After the company went out of business, Rentschler formed the Wright Aeronautical Corporation, Pratt and Whitney Aircraft Company, and United Aircraft and Transportation Corporation, which became United Airlines. He was named to the National Aviation Hall of Fame for his work in developing the Wasp jet engine, establishing the first coast-to-coast airline, and helping manufacture the first practical helicopter.

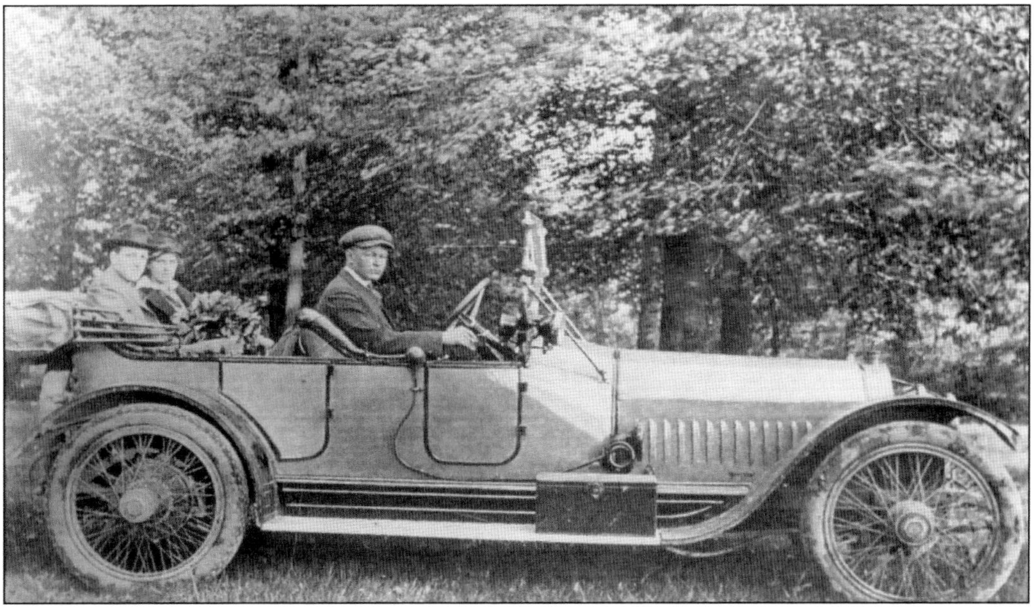

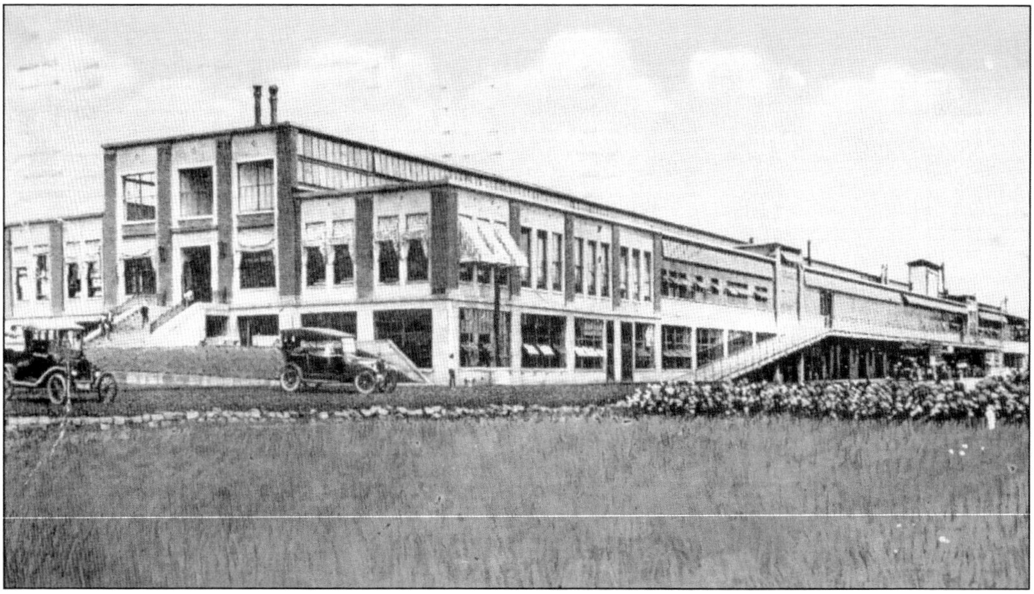

Henry Ford and Son, Inc. opened this plant in 1920 to build gasoline-powered Fordson tractors. Shortly after opening, however, the factory shifted to manufacture wheels for the Ford Model T automobile. More than 12 million wheels were made in Hamilton from 1920 to 1927, the year the plant was retooled to manufacture more than 22 million wheels for Ford's Model A automobile, before shifting to make other car parts in 1937. (Courtesy of Rob Wile.)

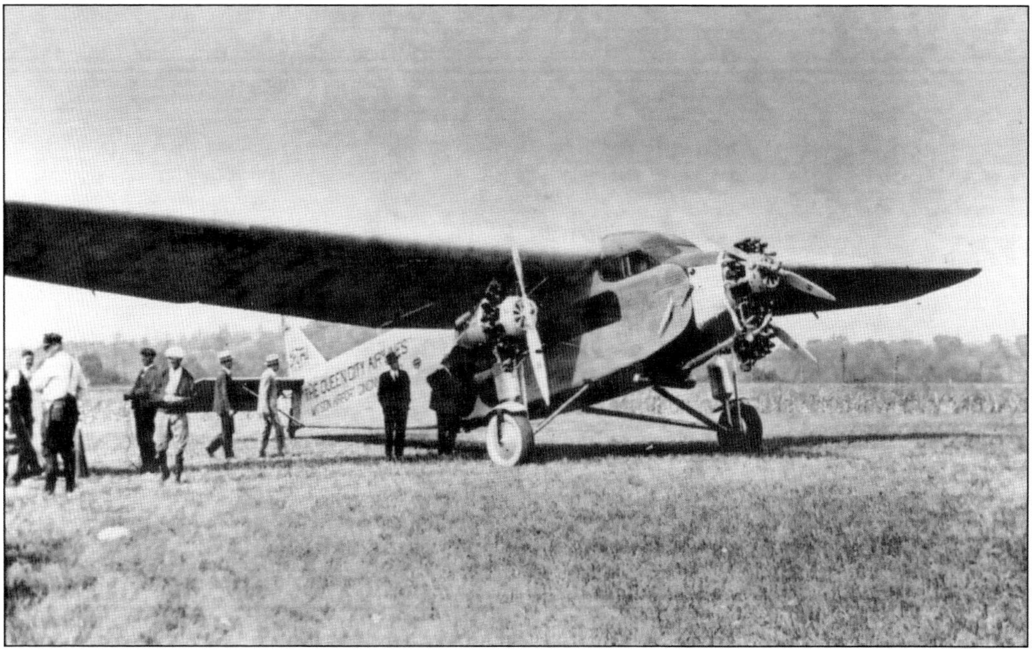

The Ford Airport was on a 500-acre tract of farmland to the north of the factory. It was in use starting in 1924. This Ford Tri-Motor Airplane landed at the airfield in 1928. The company announced that the airfield would cease operations in December 1929, but planes were still using it as late as October 1930. The land was eventually used for water wells for the city water system. (Courtesy of Susan Myers.)

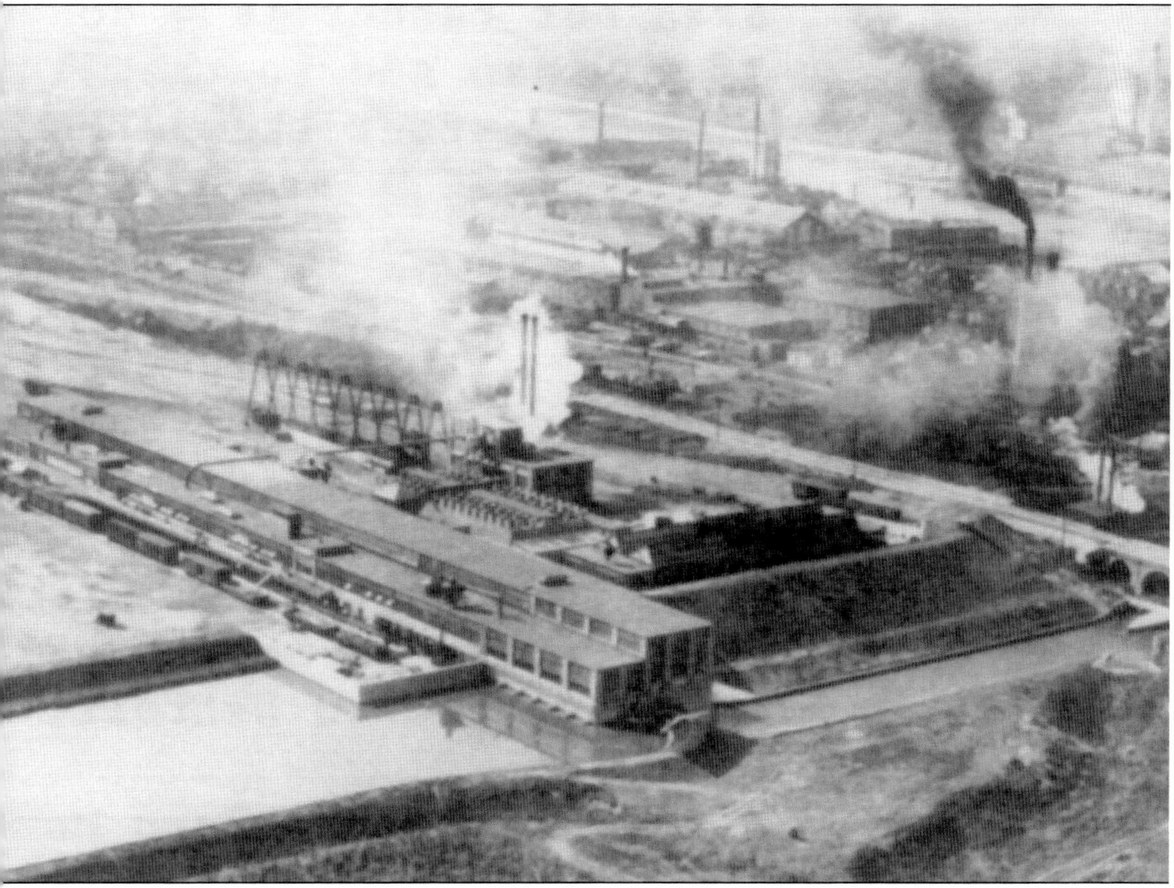

Henry Ford had purchased about 1,300 acres for the factory as well as the old Hamilton and Rossville Hydraulic Company so the company could operate a hydroelectric plant to power the factory. The plant was built where the hydraulic system's little reservoir was located. This aerial view of the factory shows the vestiges of the hydraulic and the plant's three buildings, including its two-story main building, paint shop, and boiler shop. During World War II, the plant made parts for bomber engines. The company began to shut the plant down in 1950. In 1951, the Bendix Aviation Corporation bought the factory and manufactured parts for jet engines and other vehicles until 1962. The plant was used by the Ward Manufacturing Company to make camping trailers before the Chem-Dyne Corporation opened a chemical-waste storage site on the property in 1975.

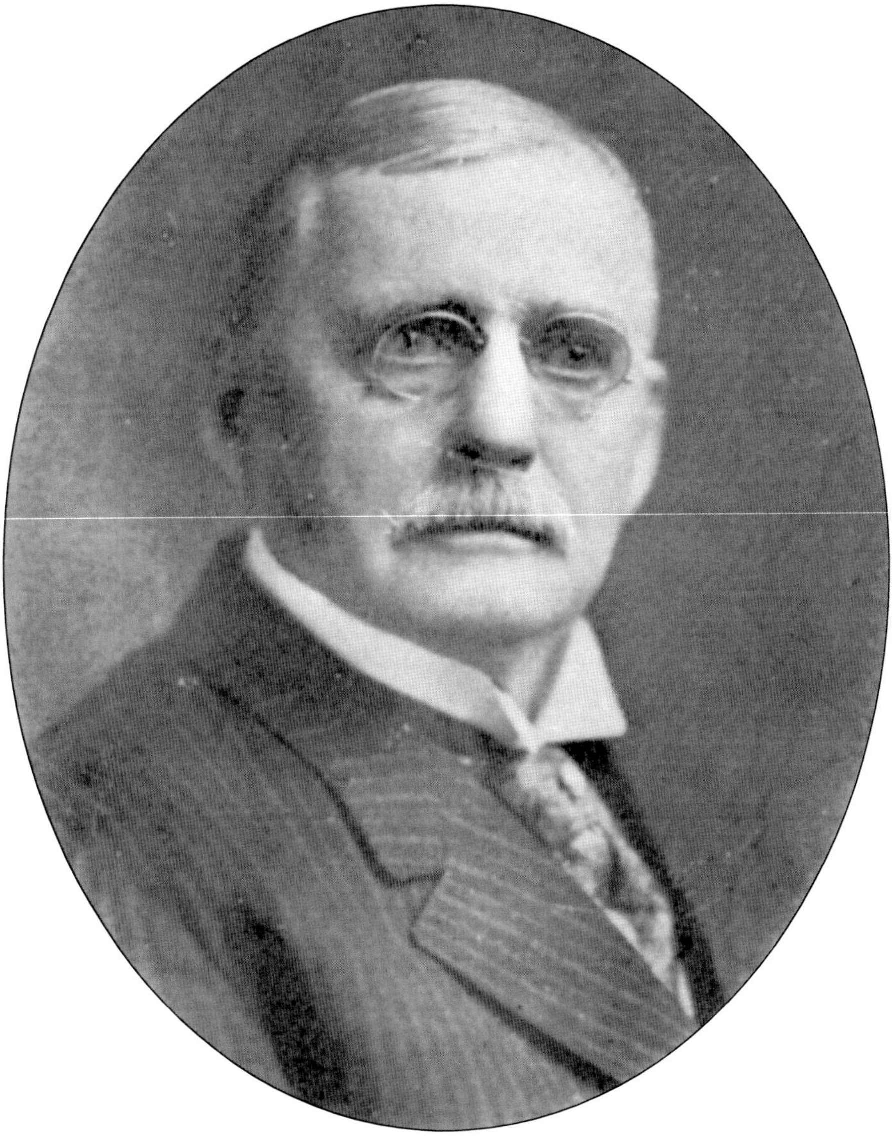

John Clinton Hooven, seen here in a photograph published in a 1914 county atlas, was a particularly innovative industrialist who played important roles in several Hamilton companies. In 1878, he became sole proprietor of the family farm-implement business and manufactured the Monarch line of portable steam engines and threshers that he had designed. When the Owens, Lane and Dyer Company went out of business in 1879, Hooven formed a group made up of James E. Campbell, George Helvey, Job Owens, George Adam Rentschler, and Henry Sohn to buy it. The new business was incorporated in 1880 as Hooven, Owens, Rentschler and continued making both of the Monarch and Eclipse lines of products. In 1883, the company manufactured its first Corliss stationary steam engine and soon specialized in making Corliss engines exclusively. Hooven was an officer or director of several businesses including American Frog and Switch Company, Hooven Automatic Typewriter Company, Dayton Rubber Company, and F.P. Stewart Granite Company. He helped finance the Cincinnati, Lawrenceburg & Aurora Traction Company and teamed with Samuel D. Fitton to lead the effort to get economical natural-gas service for Hamilton residents.

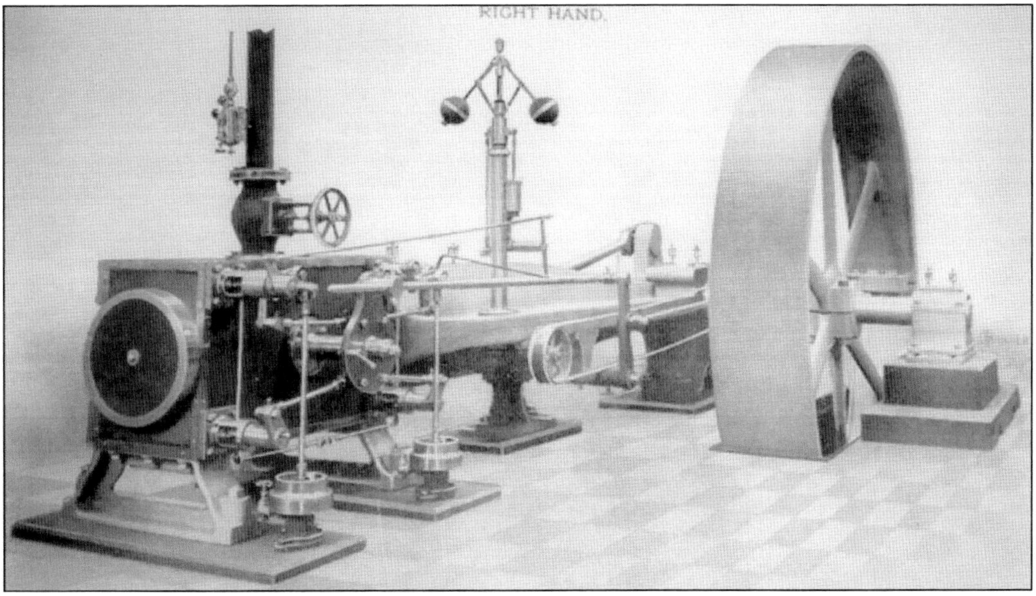

Hooven, Owens, Rentschler made this 40-horsepower Corliss steam engine in 1895. Corliss engines, patented by a Rhode Island mechanical engineer in 1849, were considerably more fuel efficient than conventional steam engines and made steam power more economical than waterpower. They also ran smoothly, maintained precise speeds, and had low maintenance costs. The company was so successful at building the engines that it became the largest exclusive Corliss engine plant in the country. (Courtesy of Richard Haid.)

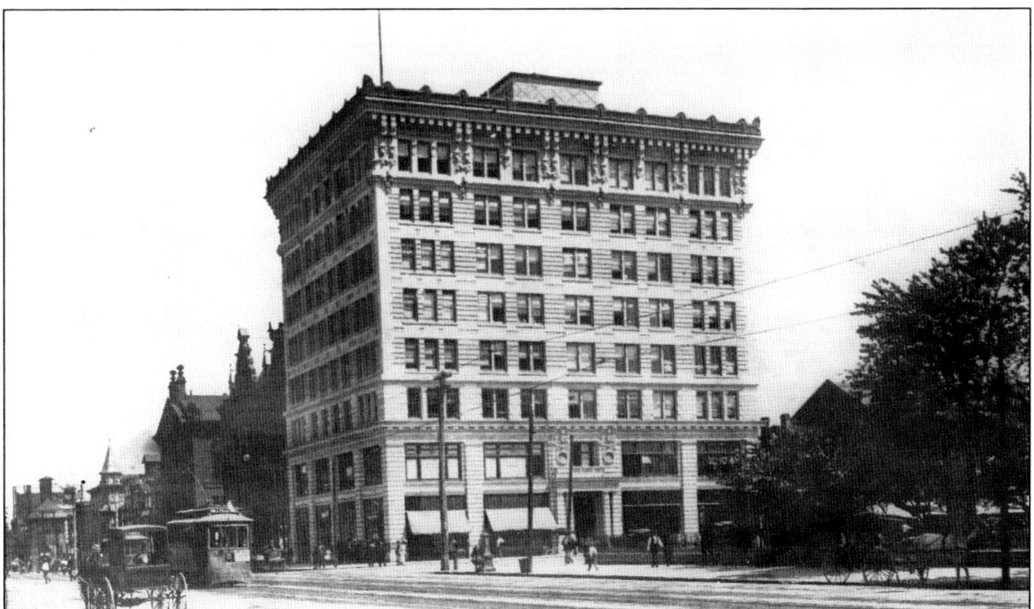

Construction of the Rentschler Building, named after George Adam Rentschler, was completed in 1906, when this photograph was taken. Rentschler was vice president of Hooven, Owens, Rentschler and president of the Hamilton Foundry and Machine Company when he opened what would be the city's tallest building until 1999. The $300,000 building would remain in the Rentschler family until 1982, when it was purchased by the Society Bank.

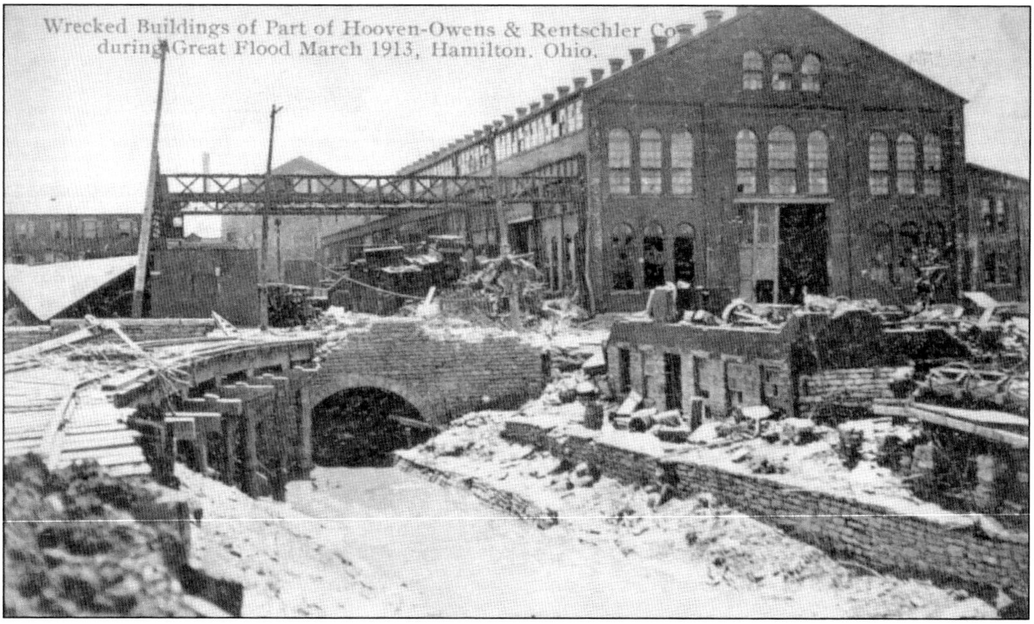

The 1913 Great Miami flood destroyed portions of the Hooven, Owens, Rentschler facilities, which had been expanded to cover several city blocks. Significant damage was also done to the tracks of the Pittsburgh, Cincinnati, Chicago & St. Louis Railroad, which served the company. The bridge arch at the center of the postcard spanned the hydraulic, which contributed to the flooding when water backed up and overflowed the channels.

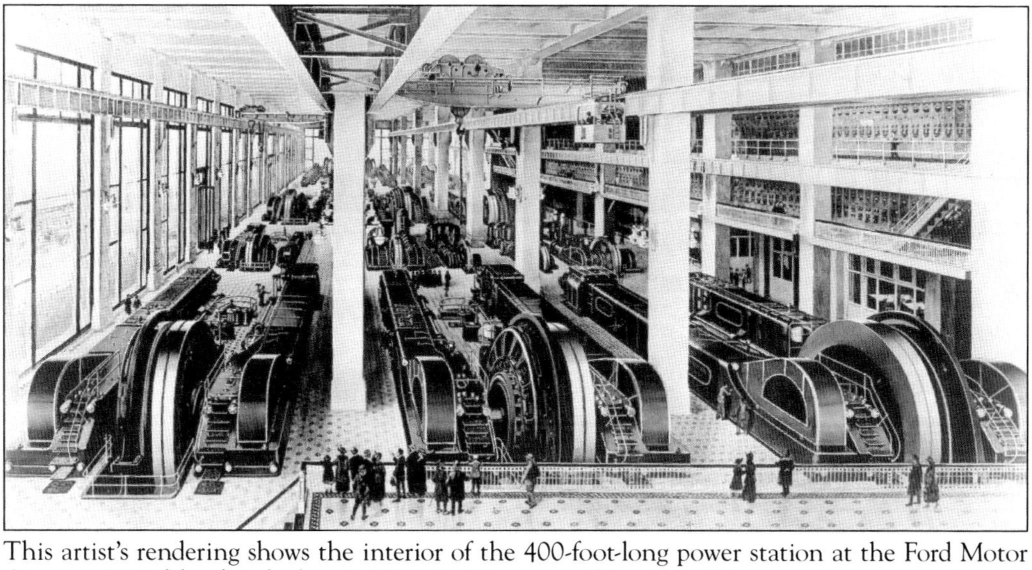

This artist's rendering shows the interior of the 400-foot-long power station at the Ford Motor Company's Highland Park plant in Detroit. Using a combination of nine enormous gas and steam Corliss engines manufactured in Hamilton by Hooven, Owens, Rentschler between 1914 and 1916, nearly 60,000 horsepower was produced to power the Ford plant.

August E. Kunker controls a 20-foot boring mill as it machines a large marine cylinder in January 1918. The apparatus was manufactured for Hooven, Owens, Rentschler by the Niles Tool Works. A white horizontal line above the third round opening from the top of the boring mill side indicates the high-water mark in the plant during the 1913 Great Miami River flood.

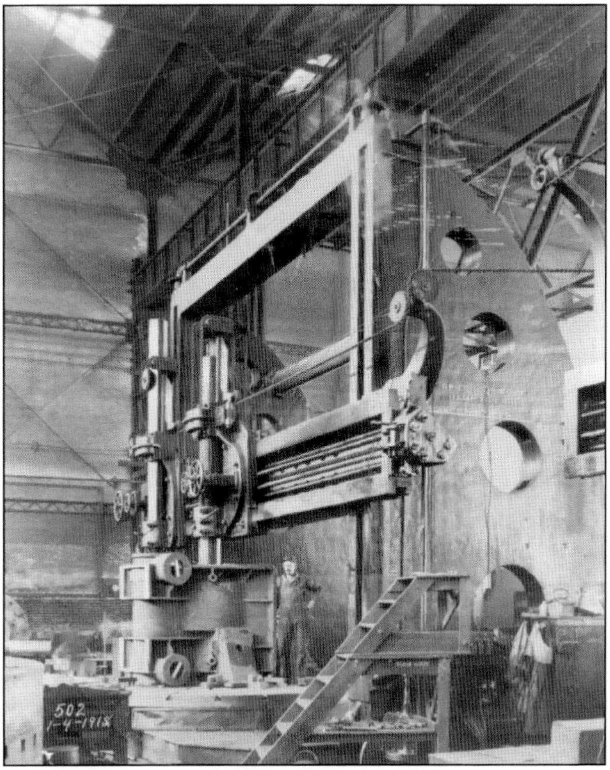

Cane-sugar mills produced by Hooven, Owens, Rentschler were sold in many tropical countries. The mill being manufactured for a Cuban plantation in this 1920 photograph had a 33-inch-by-18-inch crusher (top right) and a series of rollers to squeeze the liquid out of the sugar cane. Boilers burned the leftover stalks to produce steam for the Corliss engine that powered the mill.

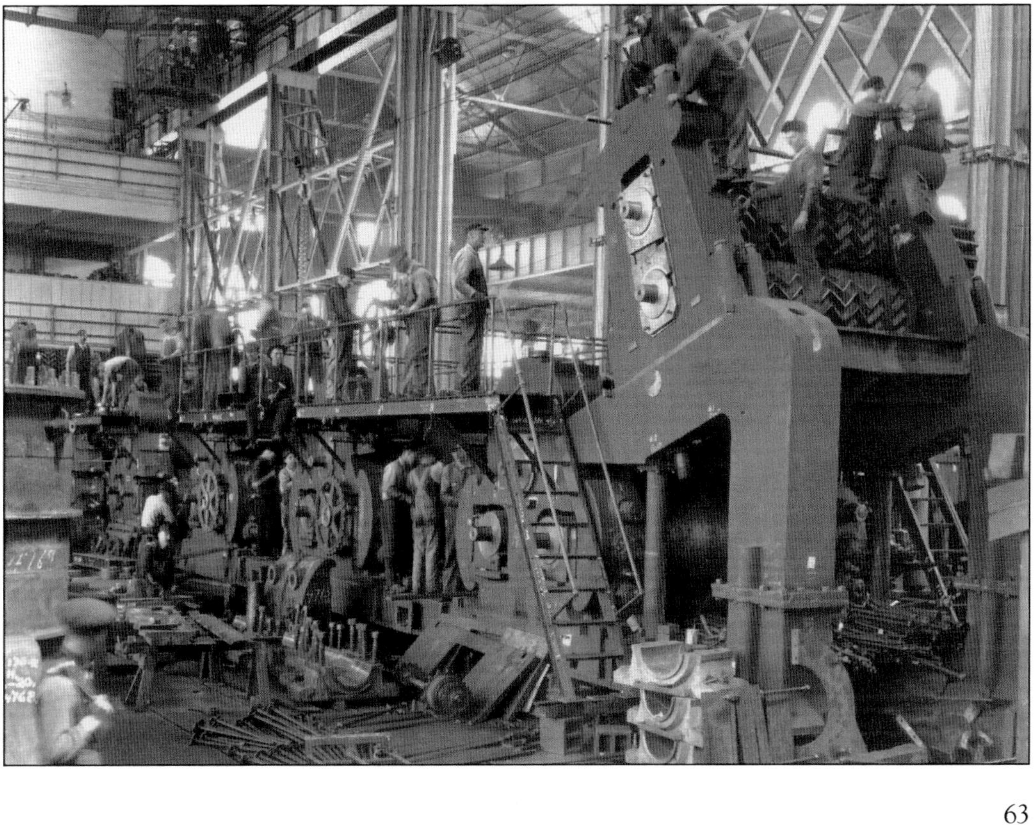

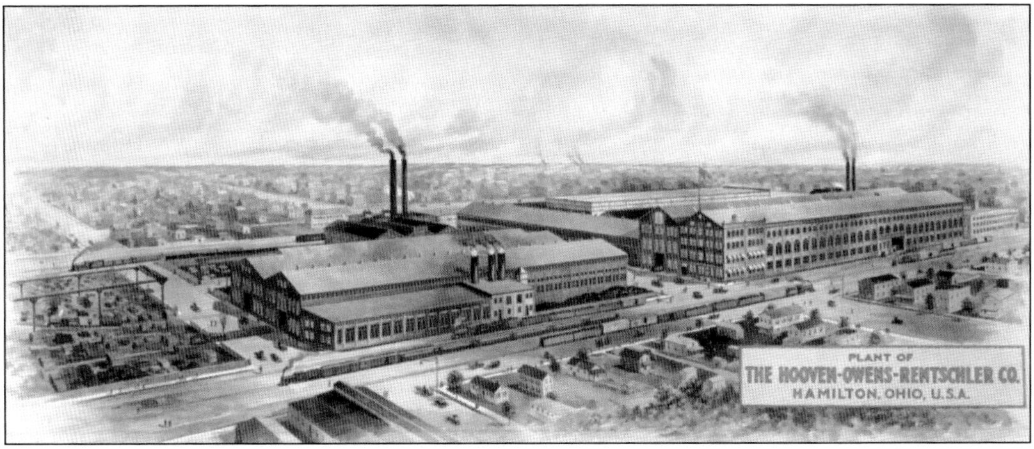

This drawing of the Hooven, Owens, Rentschler factory was published in an illustrated anniversary edition of the *Hamilton Daily News* in 1928. The plant had been expanded several times prior to 1928 and would be improved and modernized as business demands changed during the 1930s and 1940s.

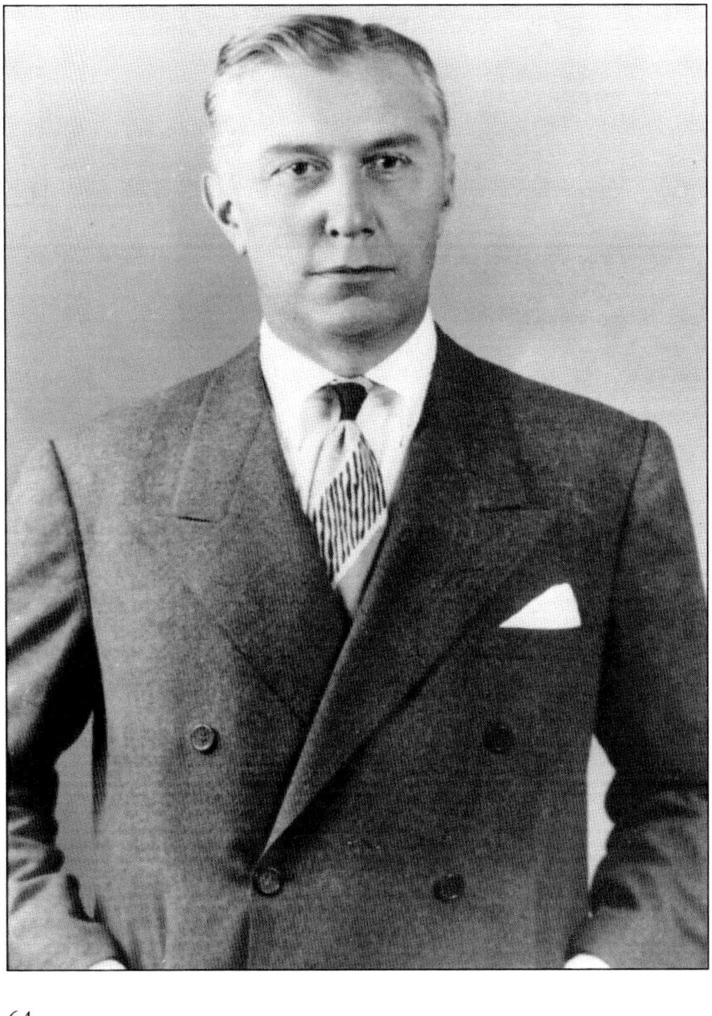

George A. Rentschler was president of Hooven, Owens, Rentschler in 1928 when his company merged with Niles Tool Works to form the General Machinery Corporation. Rentschler, a son of a founder of Hooven, Owens, Rentschler, indicated that the identities of both companies would be preserved. When General Machinery Corporation merged with Lima Locomotive in 1947 and with Baldwin Locomotive in 1950, Rentschler would remain the executive officer of those organizations.

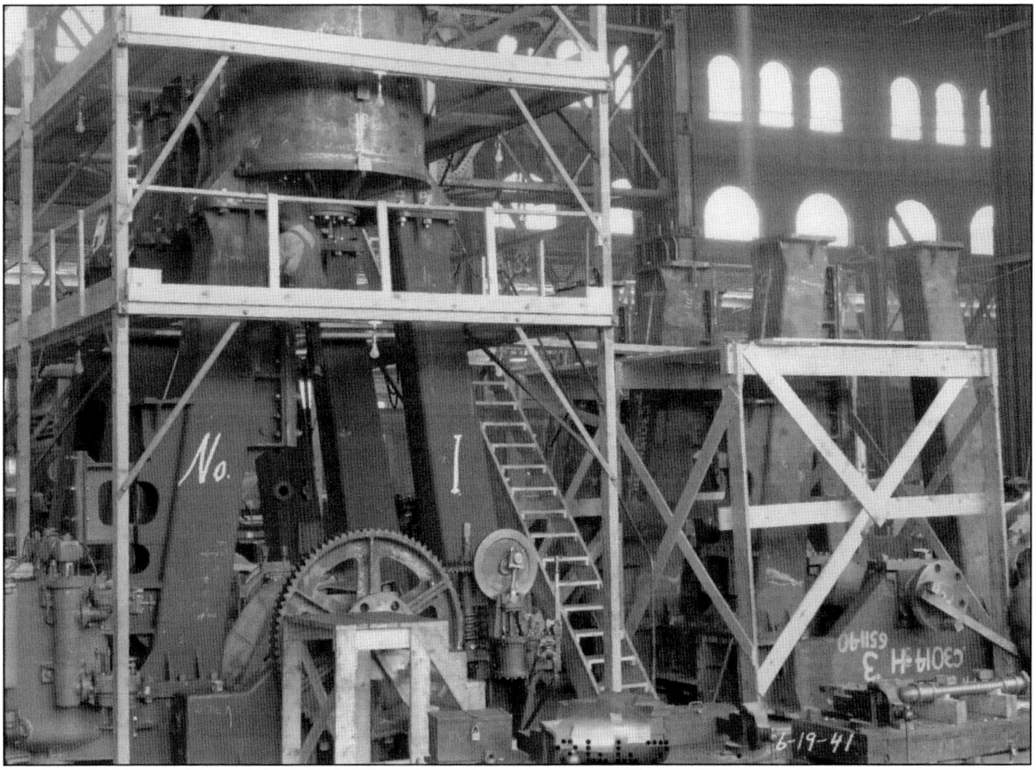

The partially completed first and third 2,500-horsepower vertical triple-expansion Liberty ship engines that Hooven, Owens, Rentschler built for the US Maritime Commission are seen above on June 19, 1941. The company was contracted to manufacture 826 of the large engines for the ships that carried food, military supplies, and troops to England. An additional 1,874 engines were built by 15 other firms across the United States. Engine No. 1 (left) was completed on July 1, 1941, and engine No. 3 was finished a short time later. Each engine weighed 287,700 pounds and was 32 feet long, 17 feet wide, and 23 feet tall. The right photograph was taken when the SS *Ocean Vanguard*, the first of the fleet of Liberty ships built for the war, was launched on August 16, 1941, at the Todd-California Shipbuilding Corporation site in Richmond, California. The ship was powered by the first Liberty ship engine built by the men and women employees of Hooven, Owens, Rentschler.

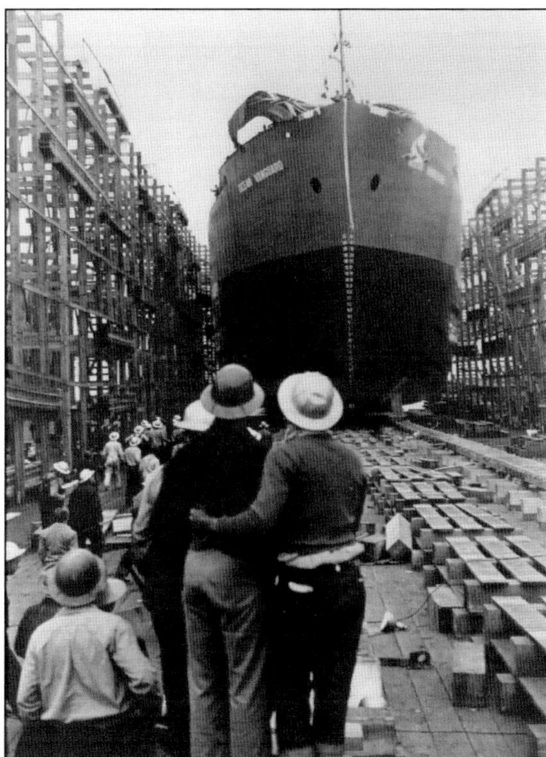

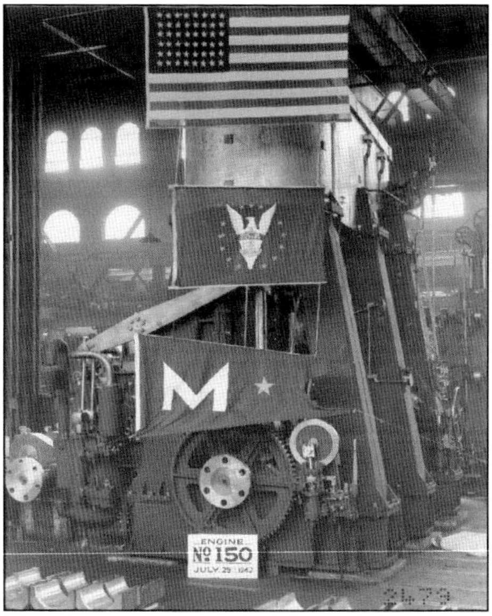

Hooven, Owens, Rentschler completed Engine No. 150 on July 29, 1942. By November 1943, the company was building the engines at a rate of one every day. The "M" flag in front of the engine indicates the Maritime Award, given to the company for its contribution to the war effort. An additional four gold stars for the flag were awarded between 1943 and 1945. The company also received the Army-Navy "E" Award and four additional white stars to go with that flag.

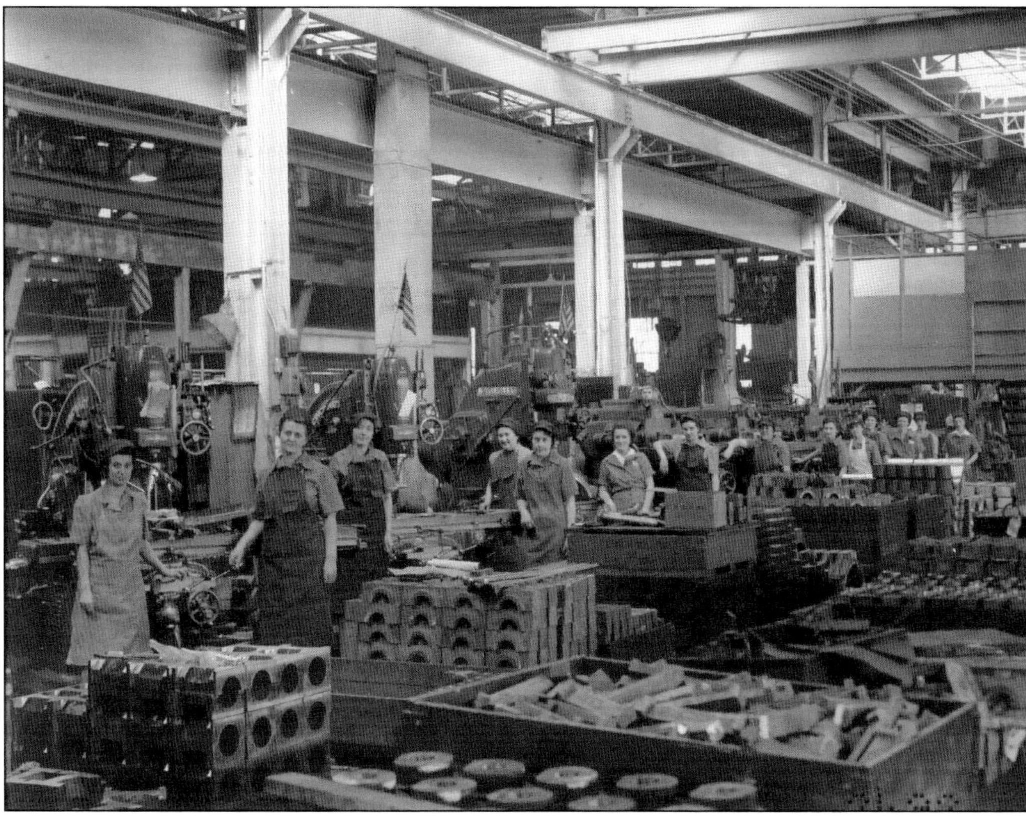

This photograph from October 1942 shows that women made up a substantial part of the workforce at Hooven, Owens, Rentschler during the war years. As many as 4,500 men and women labored at the company's plant to make artillery pieces, tanks, ship armor, Liberty ship engines, and other military hardware for the armed forces. During the war years, many company employees worked 11-hour shifts, five to seven days a week.

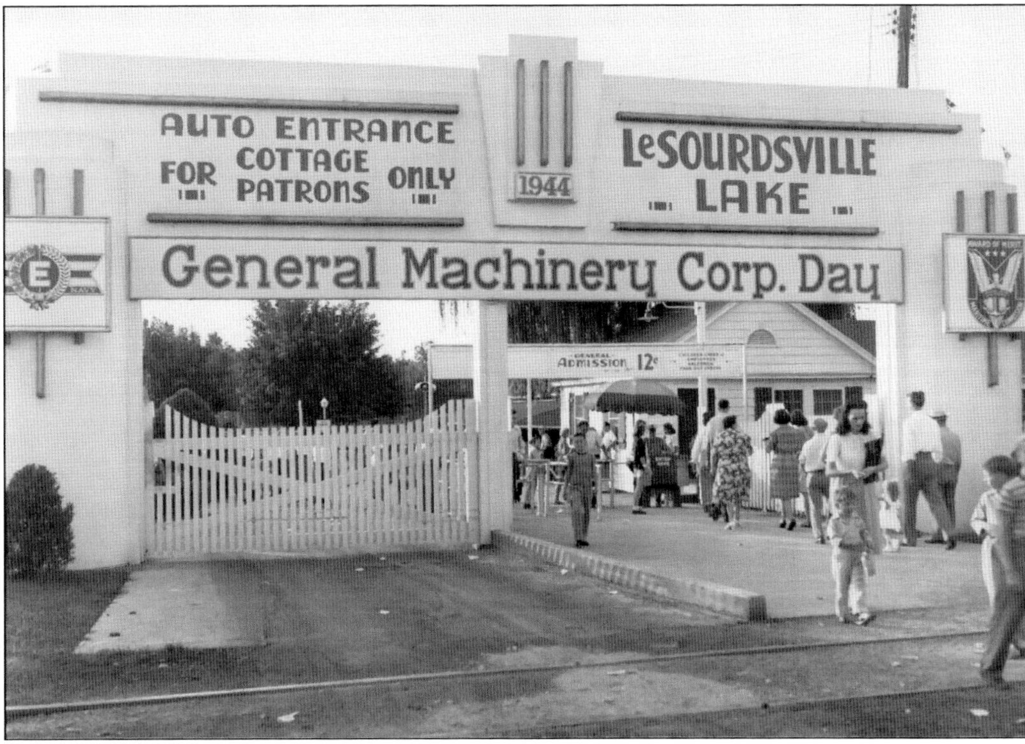

After the 1928 merger of Niles Tool Works and Hooven, Owens, Rentschler, the two companies continued to make their products under their respective brands. On August 20, 1944, General Machinery Corporation employees enjoyed a picnic at LeSourdsville Lake Park. Seen at the left of the park entrance is a reminder that employees had been awarded the Army-Navy "E" Award. On the right is a replica of the Award of Merit given by the US Maritime Commission.

Many of Hooven, Owens, Rentschler employees and their family members worked for the firm for many years. This photograph, taken at LeSourdsville Lake Park during an August 1944 company picnic, captures Bill Snider speaking to an audience after receiving a watch to honor his 50 years of service to the company. Walter A. Rentschler (left), vice president of General Machinery Corporation, made the presentation.

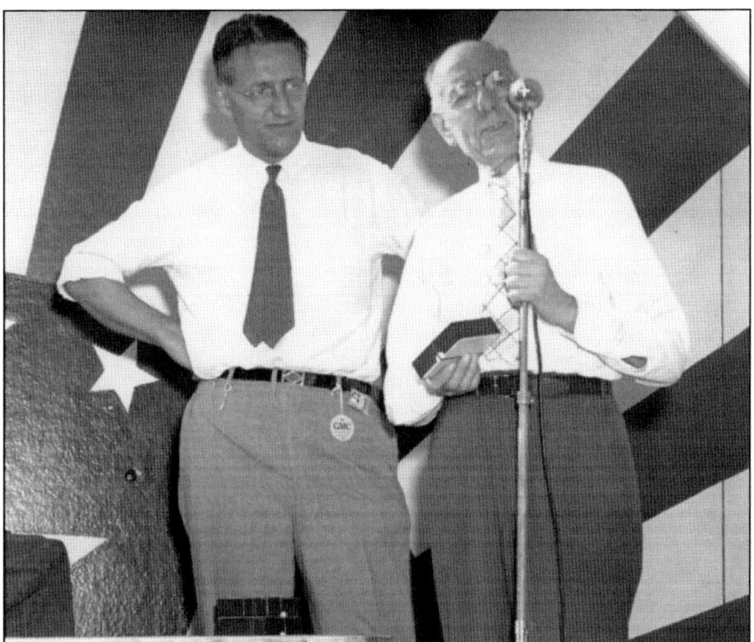

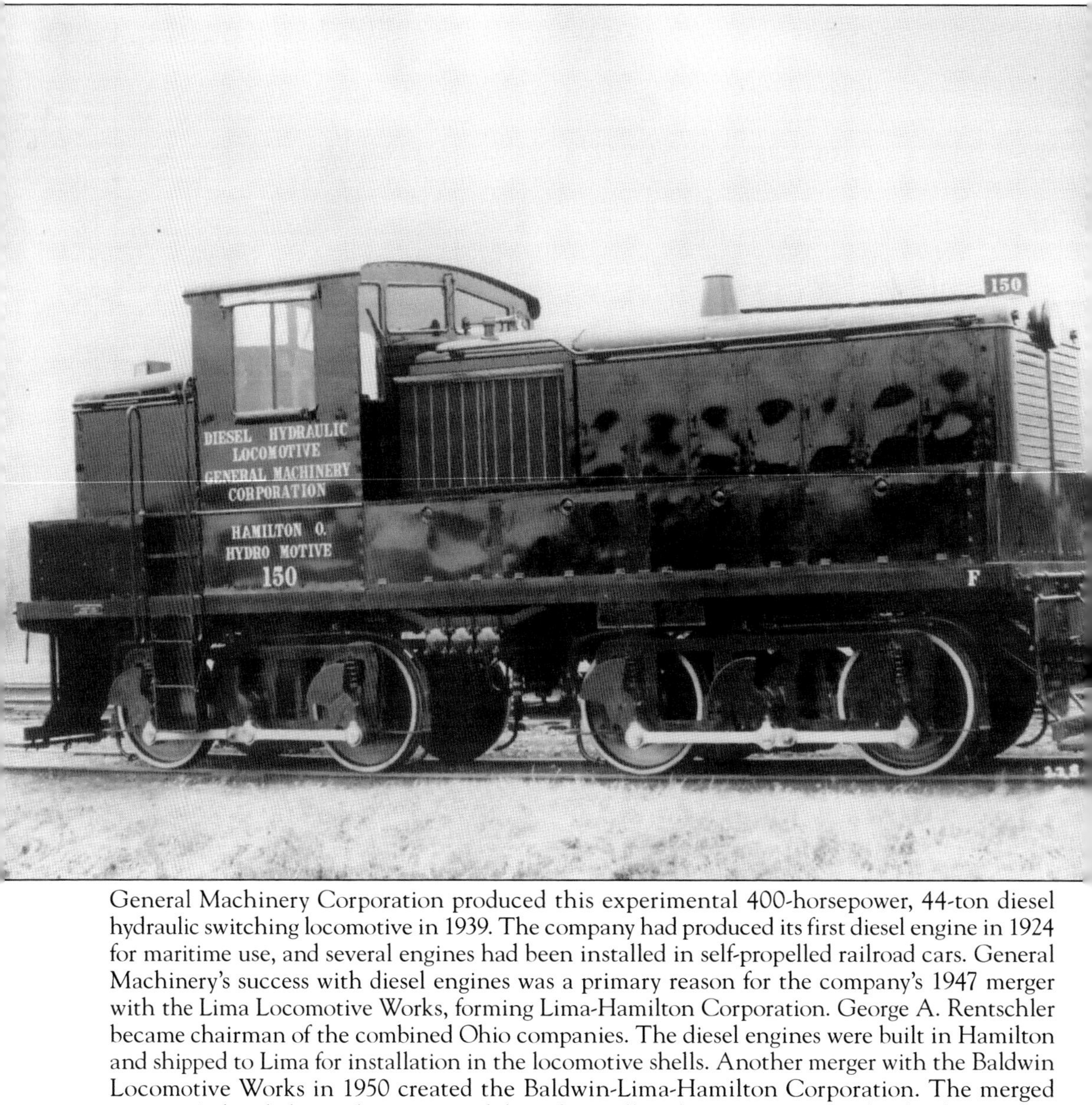

General Machinery Corporation produced this experimental 400-horsepower, 44-ton diesel hydraulic switching locomotive in 1939. The company had produced its first diesel engine in 1924 for maritime use, and several engines had been installed in self-propelled railroad cars. General Machinery's success with diesel engines was a primary reason for the company's 1947 merger with the Lima Locomotive Works, forming Lima-Hamilton Corporation. George A. Rentschler became chairman of the combined Ohio companies. The diesel engines were built in Hamilton and shipped to Lima for installation in the locomotive shells. Another merger with the Baldwin Locomotive Works in 1950 created the Baldwin-Lima-Hamilton Corporation. The merged company's diesel-electric locomotives did not fare well in the competition with other locomotive manufacturers. Baldwin-Lima-Hamilton ceased making locomotives entirely in 1956 and closed its Hamilton operations in 1960. The corporation was out of business six years later.

Five

Machine Tools and Toolmakers

Many of the first manufacturing firms in Hamilton were in the business of making implements to meet the needs of farmers. Companies operated by Charles MacBeth and David W. McClung; John Long, Peter Black, and Robert Allstatter; Thomas Millikin, J.P. Shipley, and R.B. Millikin; and several others made agricultural tools such as hay rakes, grain threshers, plows, axes, handsaws, straw cutters, planters, hoes, and other instruments. For the most part, these tools were either hand tools or used with horses providing the power to work them.

The nature of agricultural tools underwent a drastic change with the advent of the steam-powered engine. Farmers could hook their threshers, cultivators, and other forged tools to a steam engine rather than a horse to accomplish their work. Steam engines are more thoroughly discussed in chapter four.

The arrival in Hamilton of the Niles Tool Works in 1872 was the catalyst for the city to become one of America's leading centers of machine-tool production. Machine tools include lathes, planning machines, drills, shears, brake presses, slotting machines, boring machines, gear cutters, grinders, and all sorts of other tools to make machinery. Many of these tools can be quite large in size and weight in order to do the work needed to machine a tool. Often, a company needed to fabricate custom tools or modify standard ones in order to build what a customer wanted.

Niles Tool was the dominant manufacturer of machine tools in Hamilton, but there were several others that provided local workers with jobs and customers with quality products. Some of the companies featured in this chapter are the Long and Allstatter Company, which shifted from making farm implements to building punch machines, shears, break presses, and grinders; and the Hamilton Machine Tool Company, which manufactured drill presses and a variety of lathes. Other Hamilton firms presented here include the Columbia Machine Tool Company and the Hamilton Tool Company. Unfortunately, not all of Hamilton's machine-tool companies are discussed in this chapter.

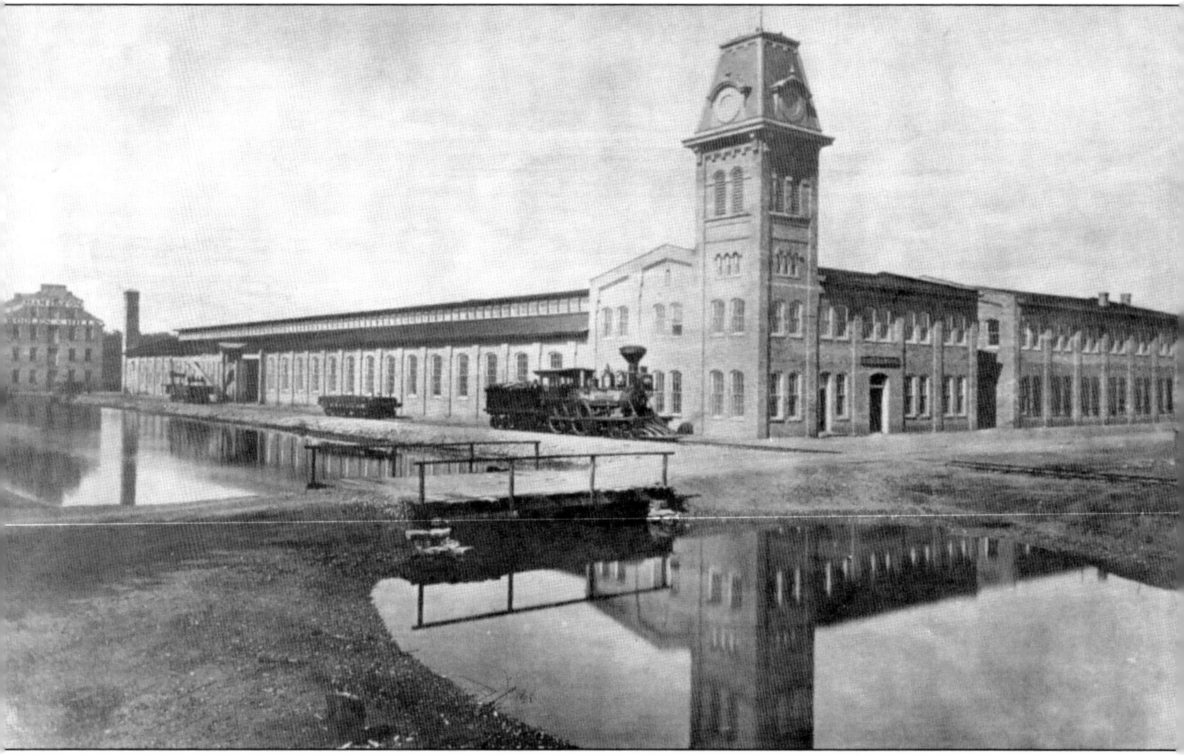

The Niles Tool Works began in Cincinnati in the early 1840s as a special department of a boatbuilding and repair business operated by James and Jonathan Niles. In 1866, George Gray, Alexander Gordon, and James Gaff bought the company and the right to do business as Niles Tool Works. Local civic leaders enticed the company to move to Hamilton by offering its owners the land, locally made bricks for a factory, and free waterpower from the hydraulic. The firm moved to this factory on North Third Street along the hydraulic in 1872. The photograph dates from early 1873, because the Hamilton Woolen Mills building (far left) was destroyed by fire on July 3. Within 15 years of its incorporation in 1874, the company greatly expanded its Hamilton facility and had established branch offices in Chicago, Philadelphia, New York, and Pittsburgh.

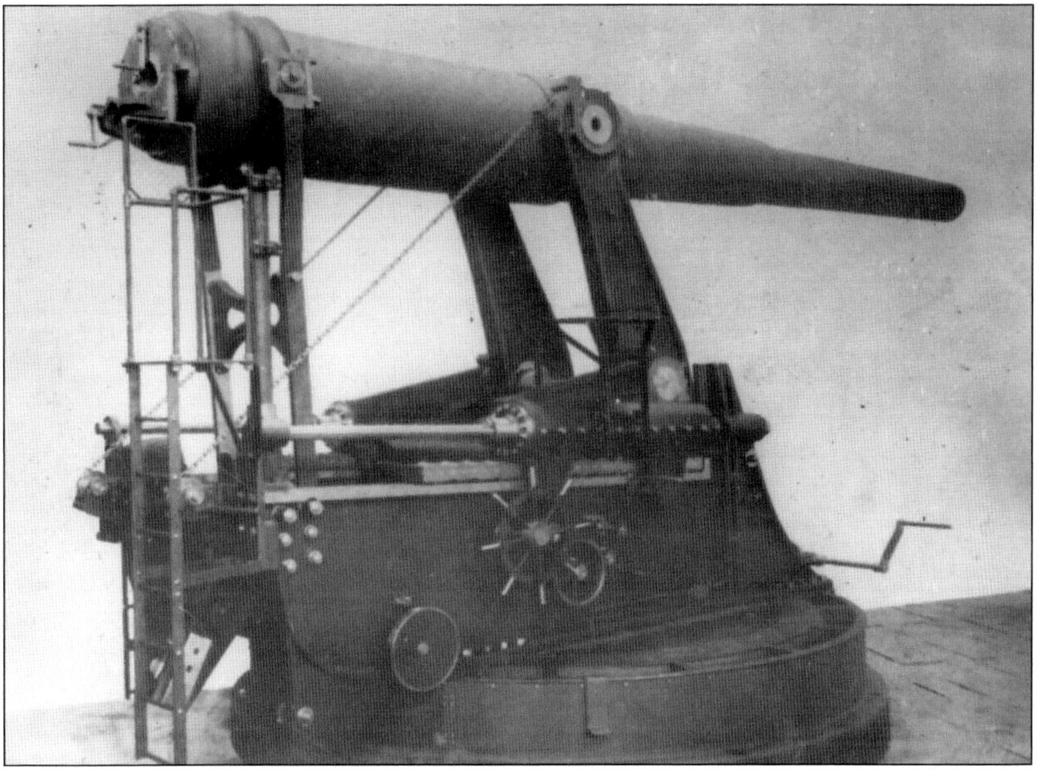

The massive gun in the above photograph was made by Niles Tool Works for use during the Spanish-American War of 1898. It was designed to be installed in US seaports for coastal defense. Each gun weighed 67,000 pounds and was 32 feet long with a 30-inch diameter breech and 16-inch muzzle. It could propel a 510-pound projectile as far as eight miles. The company also manufactured 12-foot long, 12-inch mortar guns for the war. The below photograph was taken in the Niles Tool Works gun room in 1898. Several unfinished mortar barrels are at the center of the room. Engine lathes, boring machines, and other machine tools needed to make the guns were powered by the system of pulleys and belts suspended along the walls of the room. Standing in the center of the room are Tom Flynn (left), a government inspector, and Axle Fick.

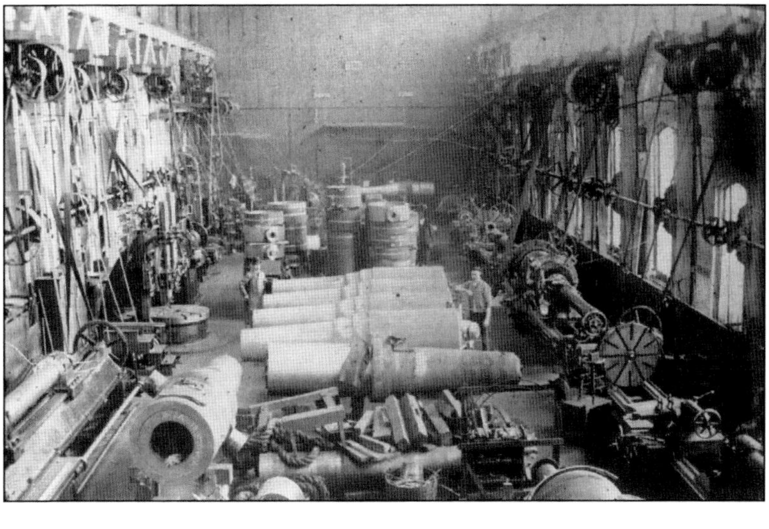

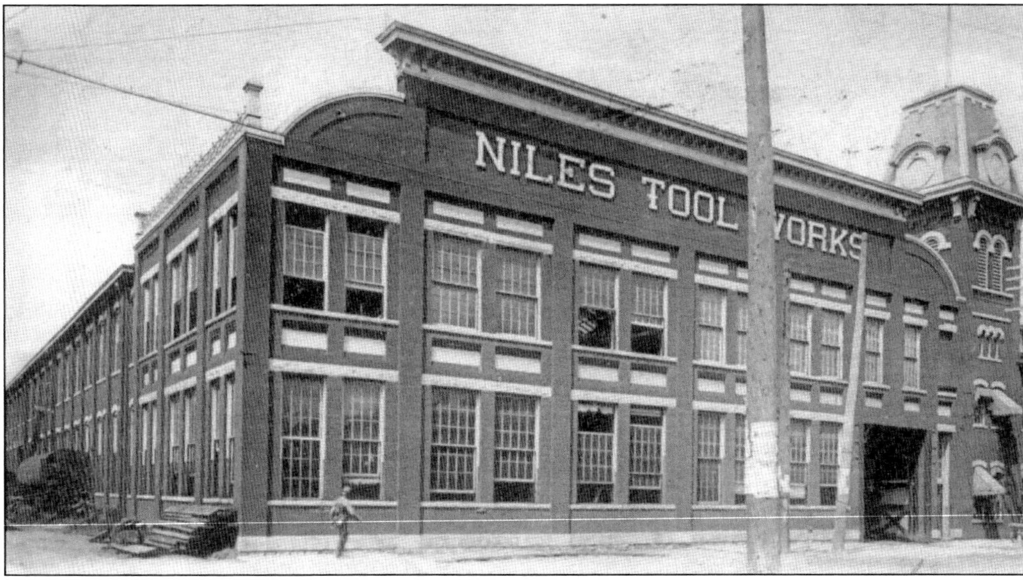

The above photograph, taken in 1905, shows that the original Niles Tool Works office and machine shop had been enlarged as early as 1887 with the addition of a second floor. A Sanborn Map & Publishing Company map of the factory dated 1887 indicates the company had a large foundry, coke ovens, machine shop, brass foundry, brass storage, storage sheds, pattern warehouse, and a railroad service through the facility. An erecting shop with two large traveling cranes was built prior to 1892. By 1899, the company had taken over the property of the Gordon Steam Pump Company on the north side of the plant, converting it into a machine shop and erecting facility. A major portion of the Louis Snider Sons paper mill on the west side was also acquired. The below photograph dates from January 1919 and was taken shortly after construction of the company's new office building was completed. The facade of the original machine shop is visible at the far right.

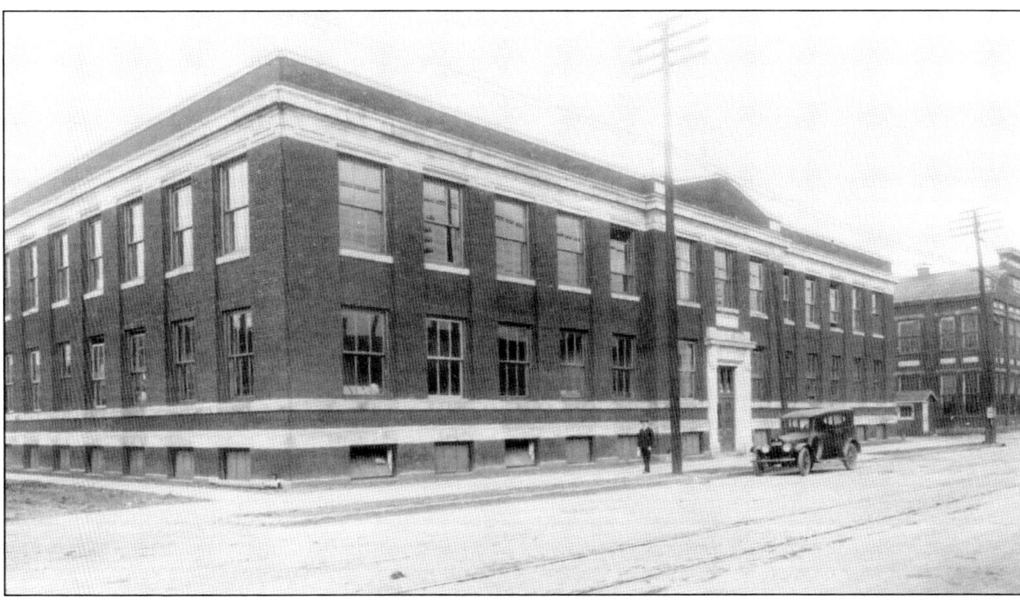

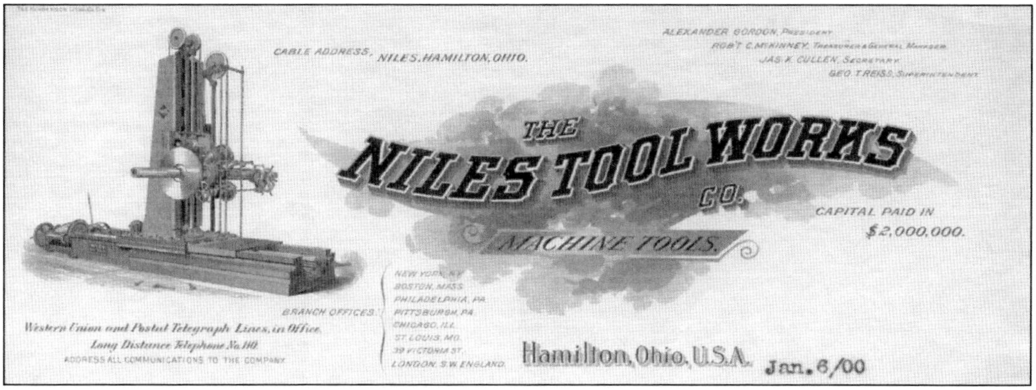
CABLE ADDRESS, NILES, HAMILTON, OHIO.

ALEXANDER GORDON, PRESIDENT
ROB'T C. MCKINNEY, TREASURER & GENERAL MANAGER
JAS. K. CULLEN, SECRETARY
GEO. T. REISS, SUPERINTENDENT

THE NILES TOOL WORKS CO.

MACHINE TOOLS.

CAPITAL PAID IN $2,000,000.

BRANCH OFFICES: NEW YORK, N.Y. BOSTON, MASS. PHILADELPHIA, PA. PITTSBURGH, PA. CHICAGO, ILL. ST. LOUIS, MO. 39 VICTORIA ST. LONDON, S.W. ENGLAND.

Western Union and Postal Telegraph Lines, in Office.
Long Distance Telephone No. 110.
ADDRESS ALL COMMUNICATIONS TO THE COMPANY.

Hamilton, Ohio, U.S.A. Jan. 6/00

By 1900, the date of the letterhead seen here, Niles Tool Works, Bement, Miles & Company, the Pond Machine Tool Company, and the Philadelphia Engineering Works consolidated and created the Niles-Bement-Pond Company. The conglomerate continued to acquire other firms, but Niles Tool Works kept its independent organization. As a result of acquisitions, expanding markets, increased international sales, and substantial domestic sales, Niles-Bement-Pond became the largest machine toolmaking company in the world. (Courtesy of Rob Wile.)

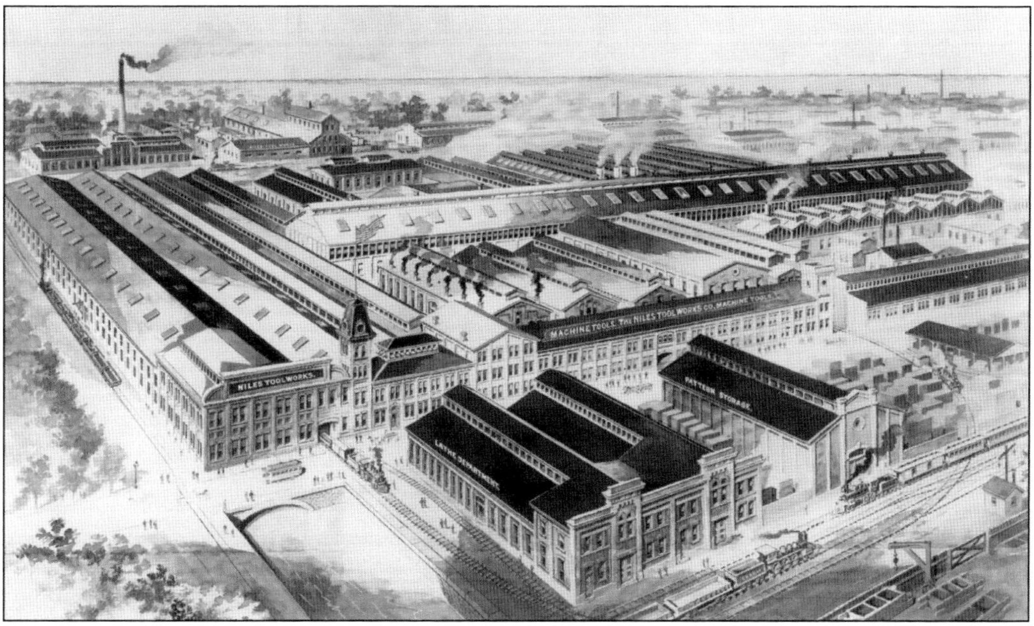

This artist's rendering provides a good image of how massive an operation the Niles Tool Works was in the early 1900s. In 1920, the Niles-Bement-Pond company's 635-page, hard-backed catalog issued in New York City offered a range of products so great that the firm could offer to build and completely equip new industrial plants with Niles-Bement-Pond tools, cranes, railway engines, track, and other materials.

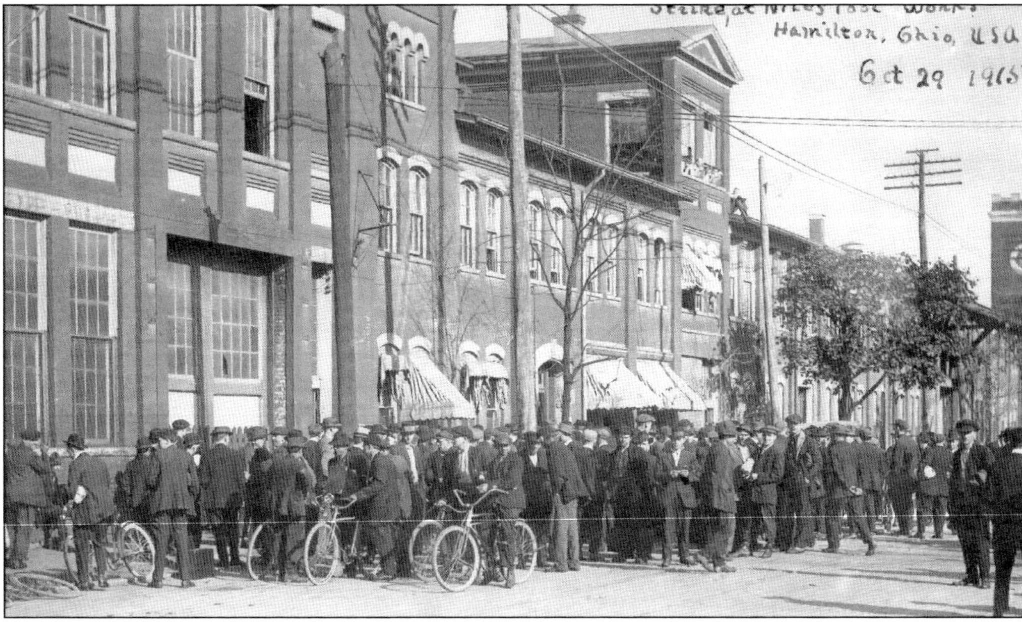

More than 620 Niles Tool Works employees went on strike on October 29, 1915, over wages and working hours. The strike lasted several months and was not limited to the Hamilton company. At the start, nearly 1,700 workers at 22 Cincinnati-area machine shops went on strike; by May 1, 1916, almost 3,700 at 113 locations were idle. The action ended when men received a 5-percent pay increase and an agreement that working shifts would be nine hours.

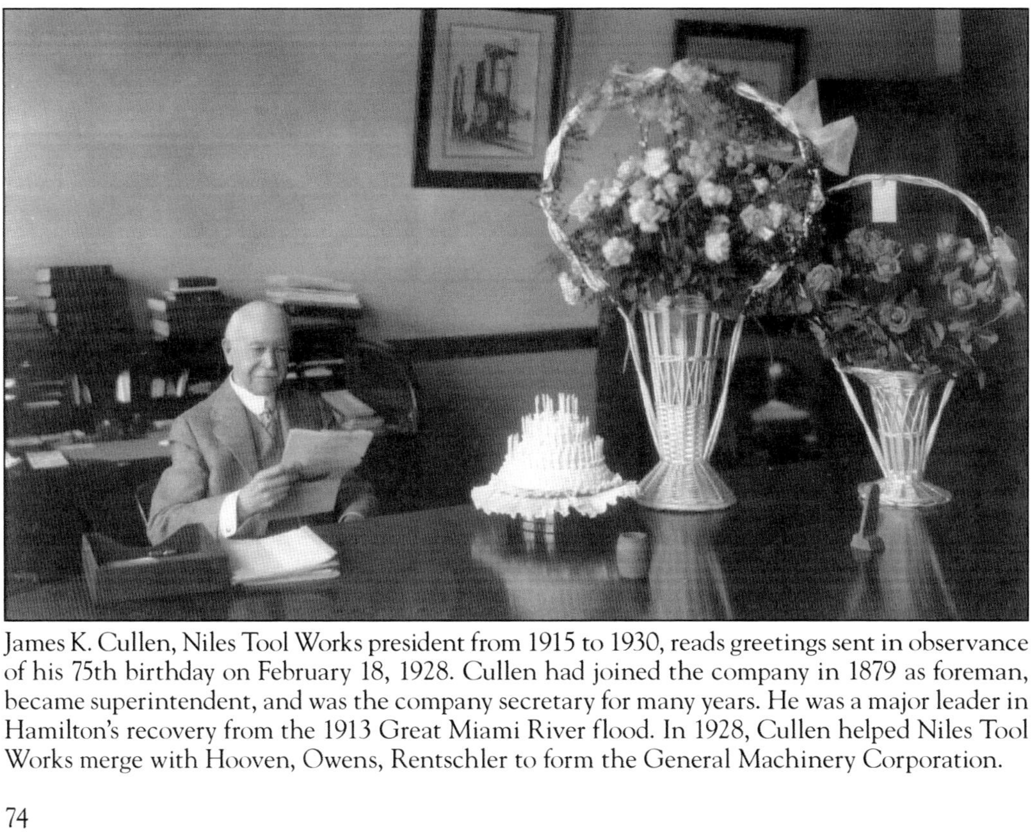

James K. Cullen, Niles Tool Works president from 1915 to 1930, reads greetings sent in observance of his 75th birthday on February 18, 1928. Cullen had joined the company in 1879 as foreman, became superintendent, and was the company secretary for many years. He was a major leader in Hamilton's recovery from the 1913 Great Miami River flood. In 1928, Cullen helped Niles Tool Works merge with Hooven, Owens, Rentschler to form the General Machinery Corporation.

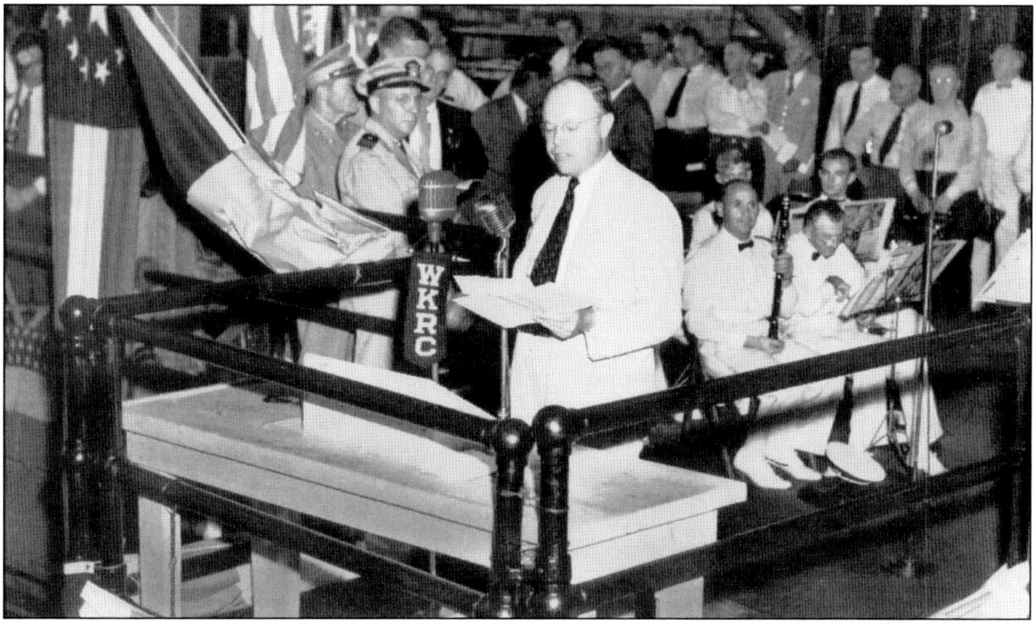

Ohio senator Robert A. Taft speaks about the significance of the Maritime Commission's "M" Award being given to the General Machinery Corporation during the ceremony on July 16, 1942. The company received the commission's Victory Flag, and all employees were given merit insignias. The commission awarded four additional gold stars during World War II. The company also received an Army-Navy "E" Award and four additional white stars during the war.

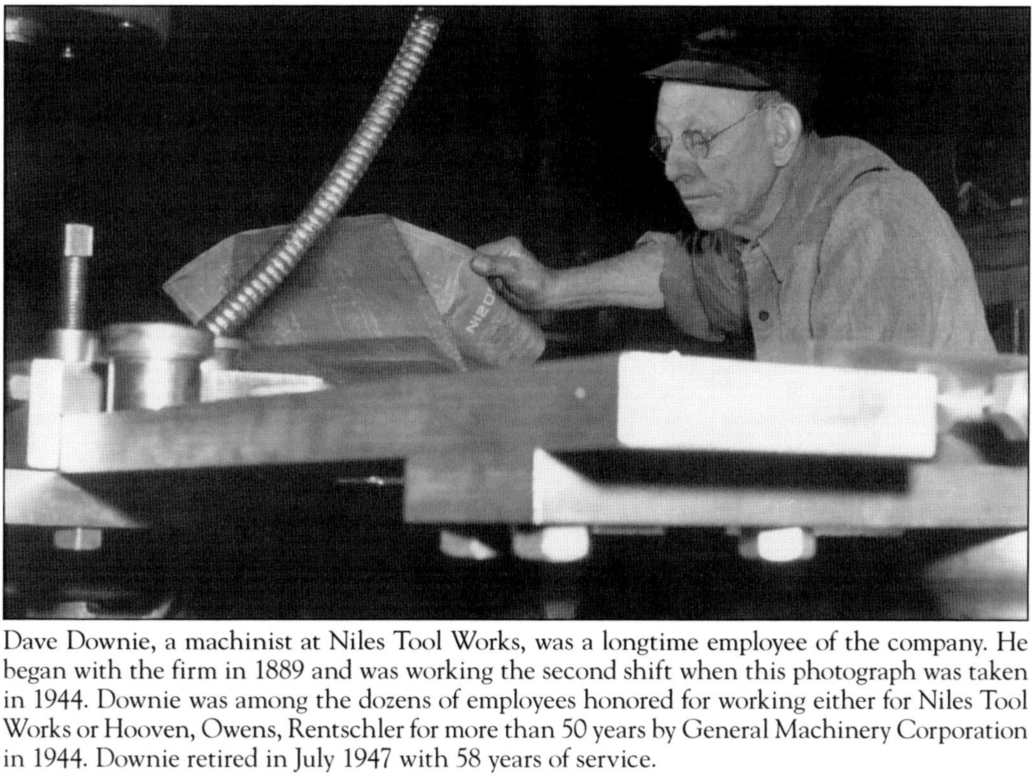

Dave Downie, a machinist at Niles Tool Works, was a longtime employee of the company. He began with the firm in 1889 and was working the second shift when this photograph was taken in 1944. Downie was among the dozens of employees honored for working either for Niles Tool Works or Hooven, Owens, Rentschler for more than 50 years by General Machinery Corporation in 1944. Downie retired in July 1947 with 58 years of service.

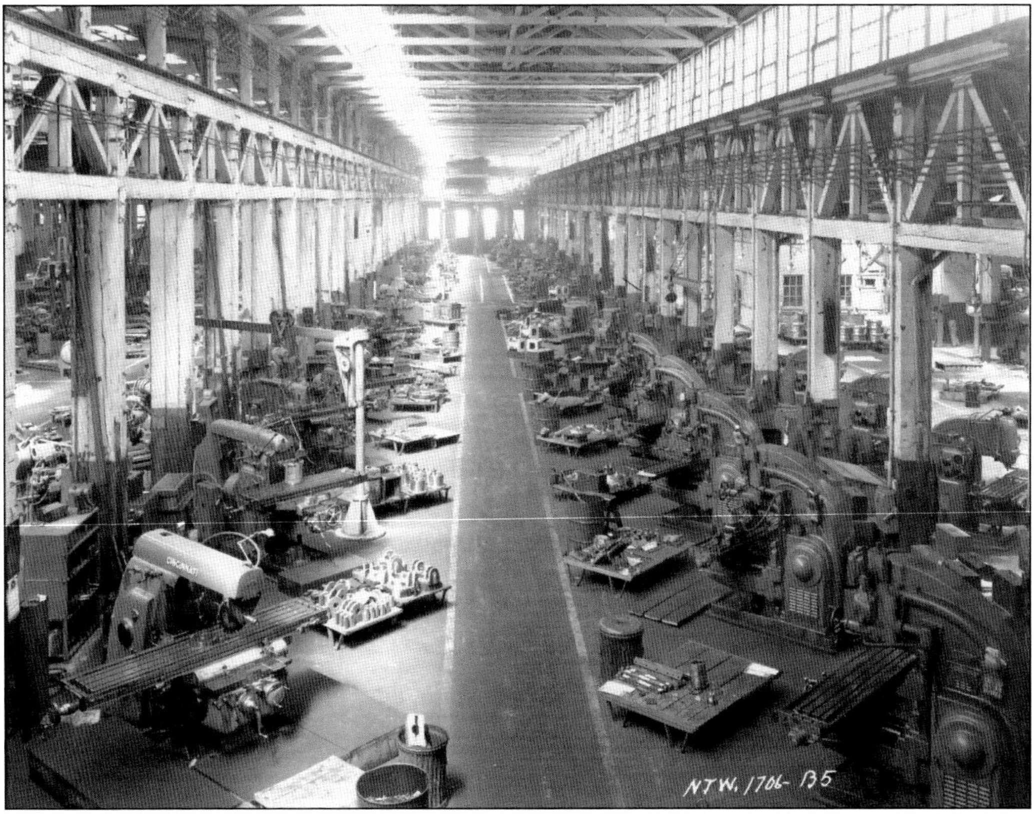

This undated photograph shows the interior of one of the Niles Tool Works machine shops. Bay No. 5 included several radial drills, grinders, and milling machines needed to tool materials. Niles Tool was part of the merger of General Machinery Corporation with Lima Locomotive Works in 1947 and then, in 1950, the merger with the Baldwin Locomotive Works. In January 1960, the company ceased operations; the history of Niles Tool Works ended.

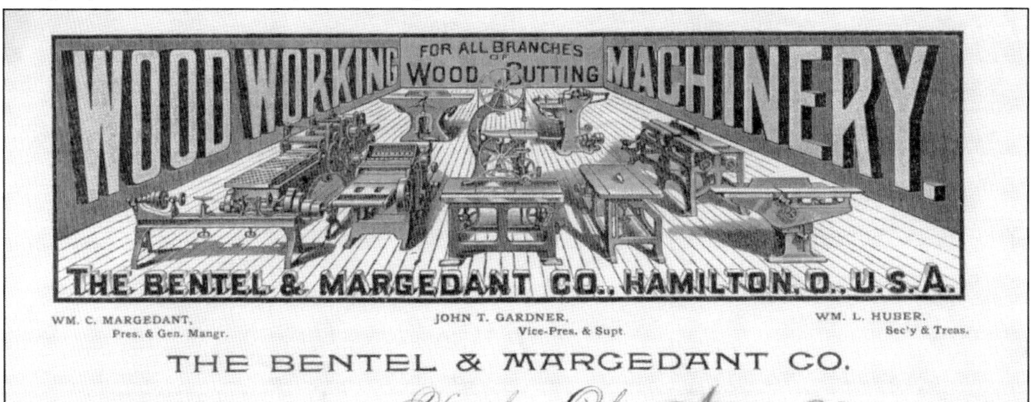

The Bentel & Margedant Company's roots go back to 1864, when the business began as Macbeth & McClung and made a variety of farm implements. When the firm was incorporated in 1885, it manufactured all sorts of woodworking machinery, including saws, routers, drills, mortises, planers, and the very popular "universal wood worker." The company's 1890 letterhead included images of several products the firm manufactured. (Courtesy of Rob Wile.)

William C. Margedant was a popular Civil War veteran who organized the centennial celebration in 1876 and the Hamilton centennial in 1891. After serving as manager since 1885, he became president of Bentel & Margedant in 1889. Captain Margedant held patents on jointers, wood planers, molders, scroll saws, band saws, drill presses, sanding machines, and wheelwrights. He headed the company until his death in 1900.

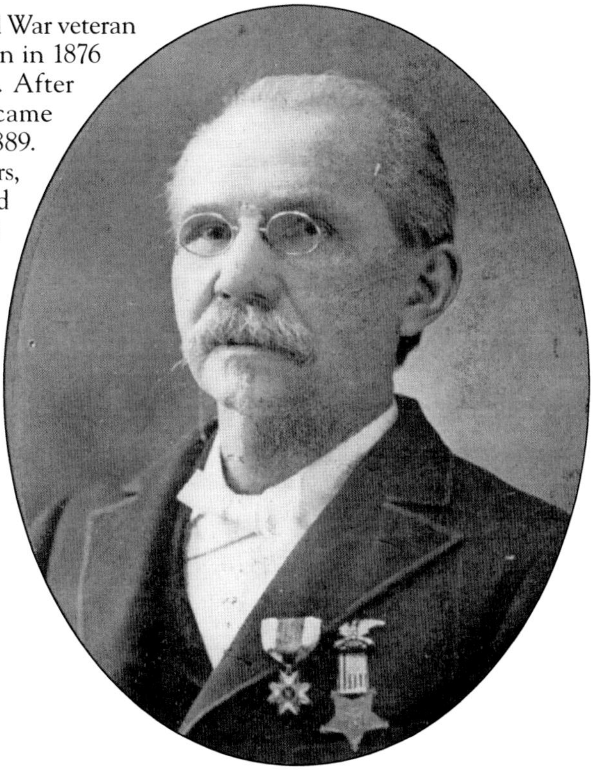

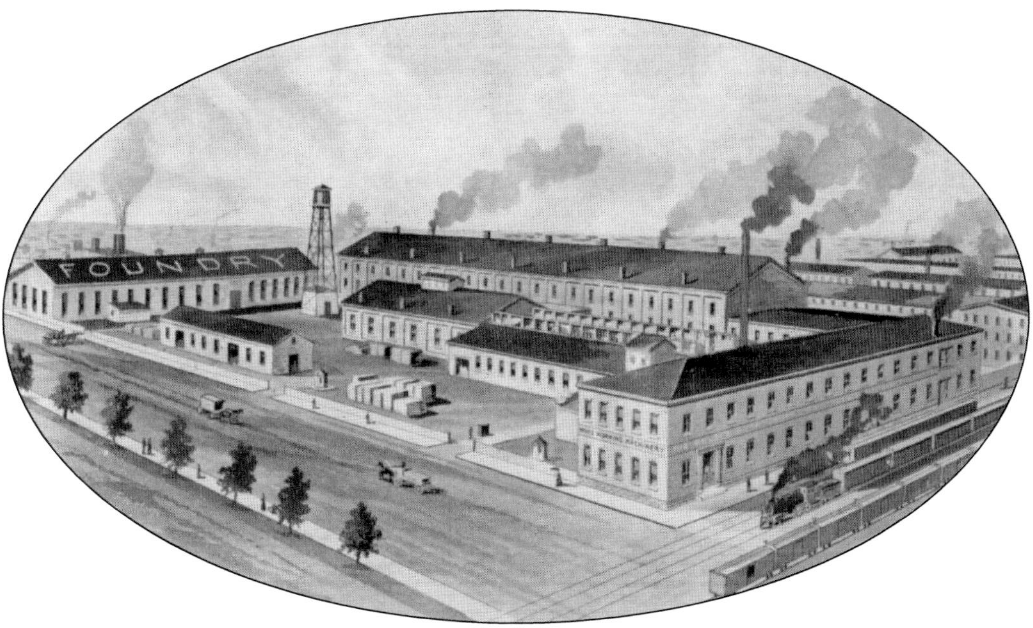

The Bentel & Margedant Company factory was located on the hydraulic near Fourth Street. This drawing of the firm's plant was copied from a 1909 newspaper supplement. In 1918, the factory was sold to the Long and Allstatter Company, and the woodworking machinery company went out of business. Members of the Margedant family then worked to build Central Foundry, a gray-iron casting company started in 1906 by Carl E. Margedant.

The business that became the Long and Allstatter Company was started in 1856 by Peter Black, John Long, and Robert Allstatter. The firm manufactured farm implements such as sickles, iron harvesters, horse rakes, plows, and cultivators like the ones on the cover of this 1890 catalog. Black sold his share of the company in 1870 and became one of the founders of the Black-Clawson Company. The Long and Allstatter Company was incorporated in 1878.

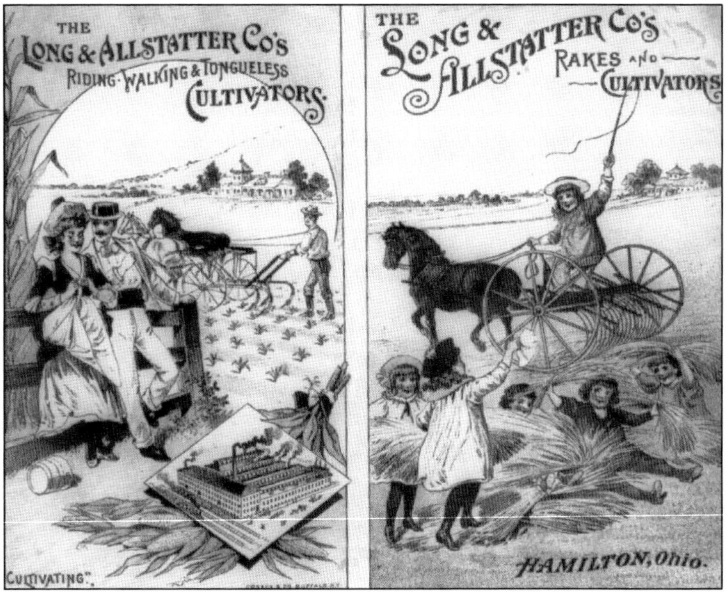

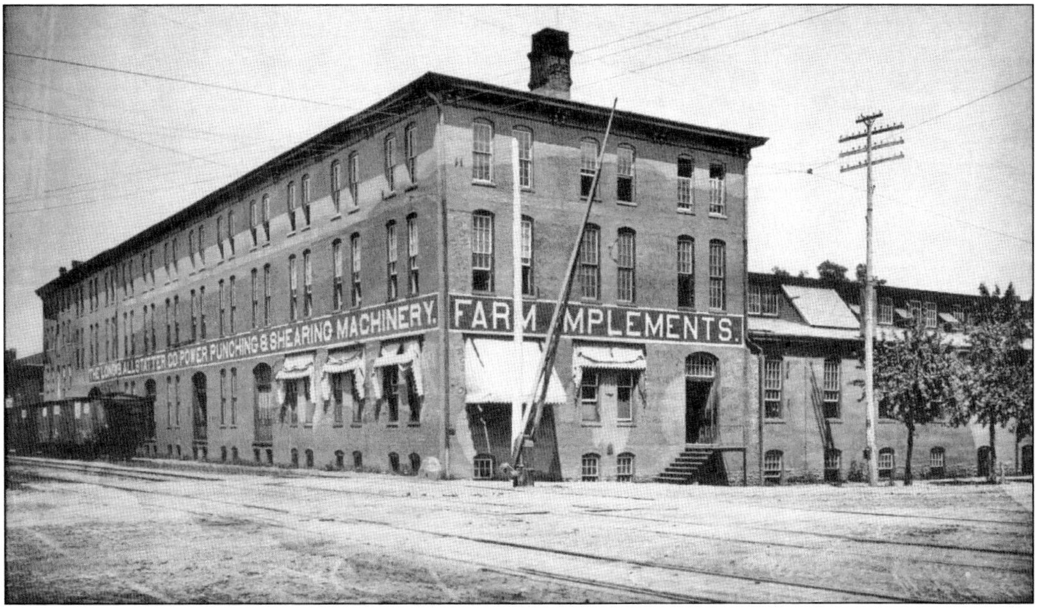

In 1871, Long and Allstatter began building the new factory on the northeast corner of High and Fourth Streets, seen here in 1893. The company ceased making its line of iron harvesters in 1873, when the new factory was completed and began manufacturing harrows, plows, and cultivators. The firm also increased its production of power-punching and shearing machines.

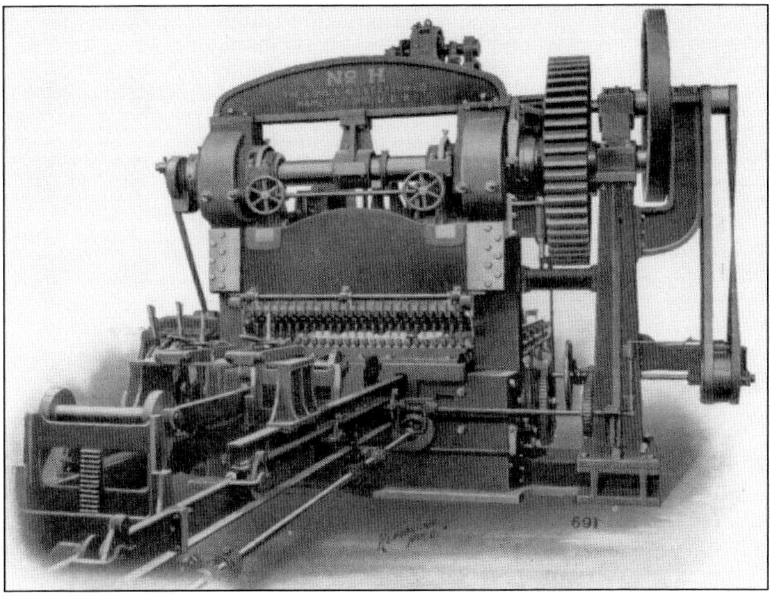

Long and Allstatter had great success with the company's line of machinery for punching, shearing, bending, straightening, and forming metal. These machines were used by the government in shipbuilding and by large rolling and steel mills around the world, safe makers, and car manufacturers. The company's multiple punch machine (above) was 86 inches wide and capable of punching 25 holes, one-inch-wide, through metal thicker than one-half inch in one stroke. It could automatically feed sheet metal over 60 feet in increments of one-sixteenth of an inch. The below photograph shows one of Long and Allstatter's gate shears. This machine could trim or split long sheets of metal from one-quarter-inch to almost two-inches thick. By 1935, the company was under the control of a bankruptcy trustee. In 1939, the Long and Allstatter factory was occupied by the Columbia Machine Tool Company.

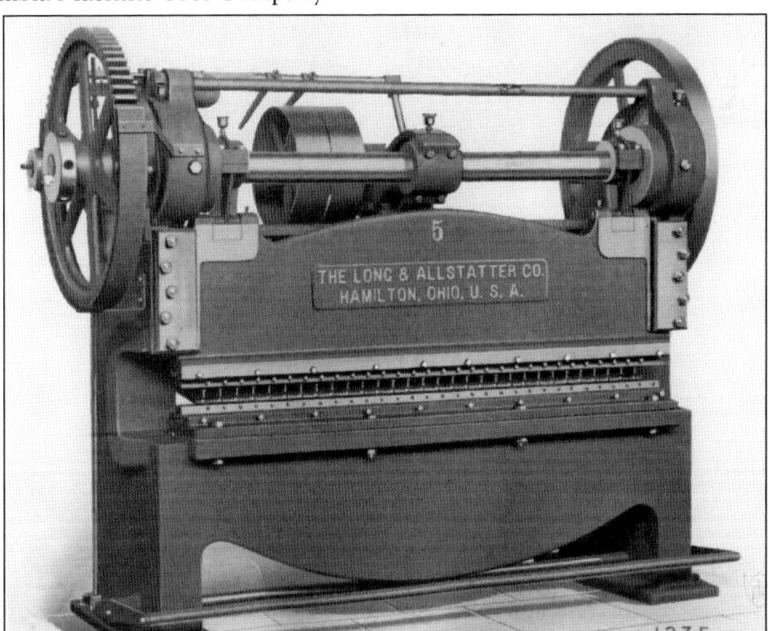

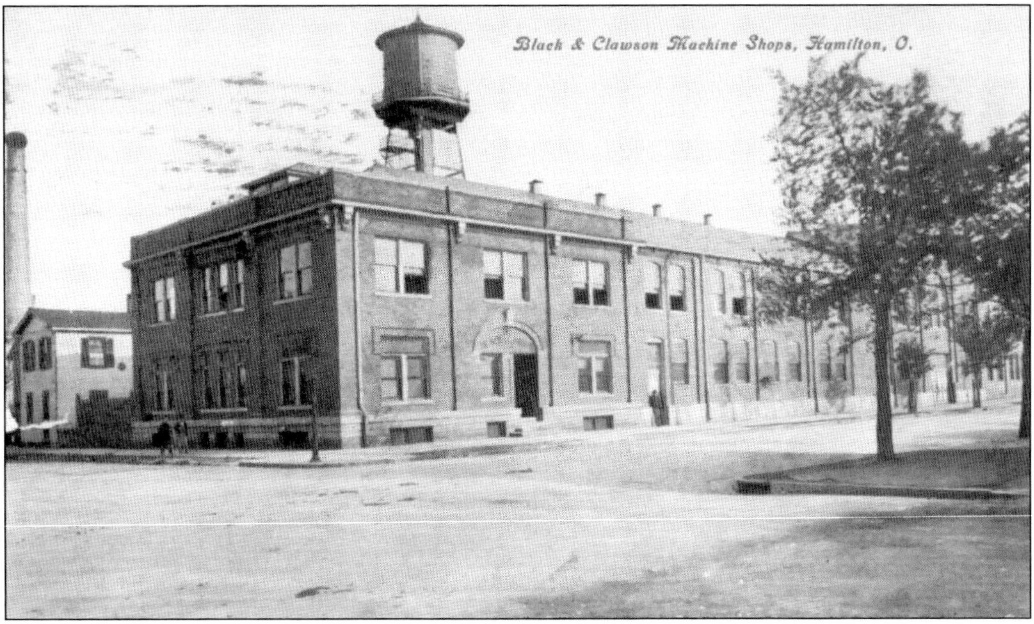

The Black-Clawson Company, formed in 1874, began manufacturing paper-mill machinery in 1878 at this plant located along the hydraulic at North Second and Vine Streets. The business was incorporated in 1883; by 1908, when this postcard photograph was taken, it was one of the largest paper-mill machinery makers in America. The company's products included paper and pulp machines, paper cutters, inking mills, drying machines, coolers, and evaporators.

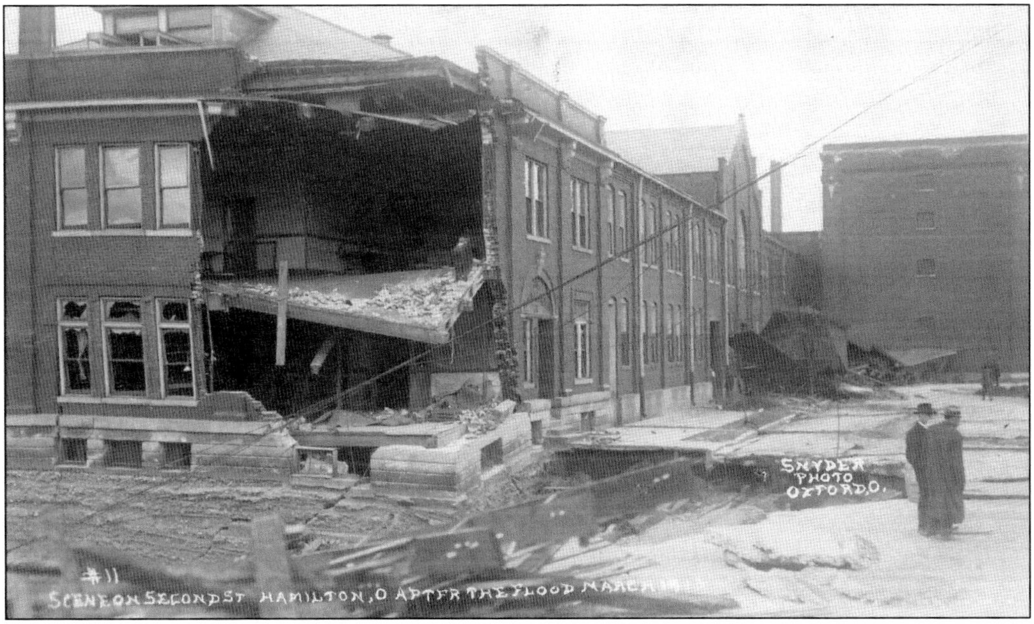

A portion of Black-Clawson's office building was damaged during the 1913 Great Miami River flood. The firm rebuilt the plant and continued exporting its Fourdrinier papermaking machinery and other related products to Great Britain, Europe, Canada, Australia, South America, and Japan. The company also had success with its development of a cylinder machine, seamless dryer, and triple-deck dryer arrangement.

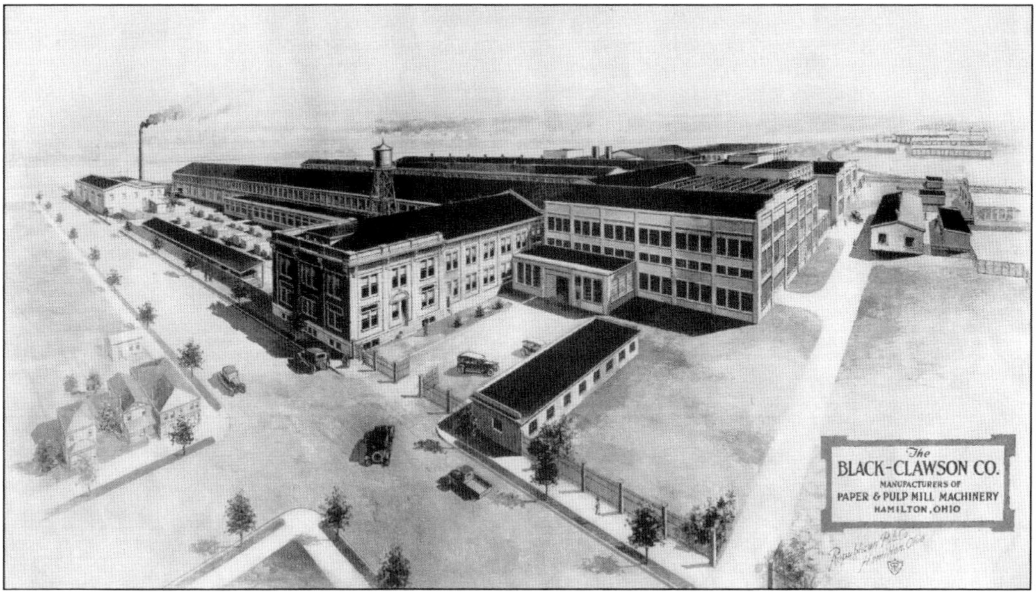

In 1926, two years before this artist's rendering of the local plant was made, Black-Clawson purchased the Shartle Brothers machine shop in Middletown. The company bought the Dilts Machine Works of Fulton, New York, in 1940, and Bagley and Sewall Company of Watertown, New York, in 1954. The company moved into recycling and, by 1997, the year Black-Clawson was absorbed by Thermo Fibertek, more than 80 percent of America's recycled paper was produced on Black-Clawson equipment.

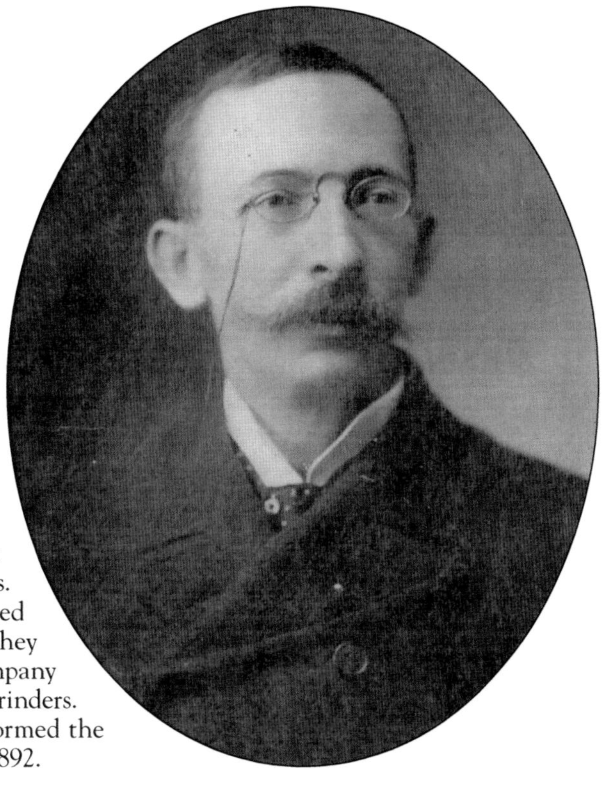

Charles F. Hilker began his National Car Seal Company in 1890, in a factory located on the hydraulic at the northeast corner of Monument and Market Streets. The firm made metal strips and seals used by railroads to lock freight cars after they were loaded for transit. In 1891, his company began making drill presses and emery grinders. Hilker sold the car-seal business and formed the Hamilton Machine Tool Company in 1892.

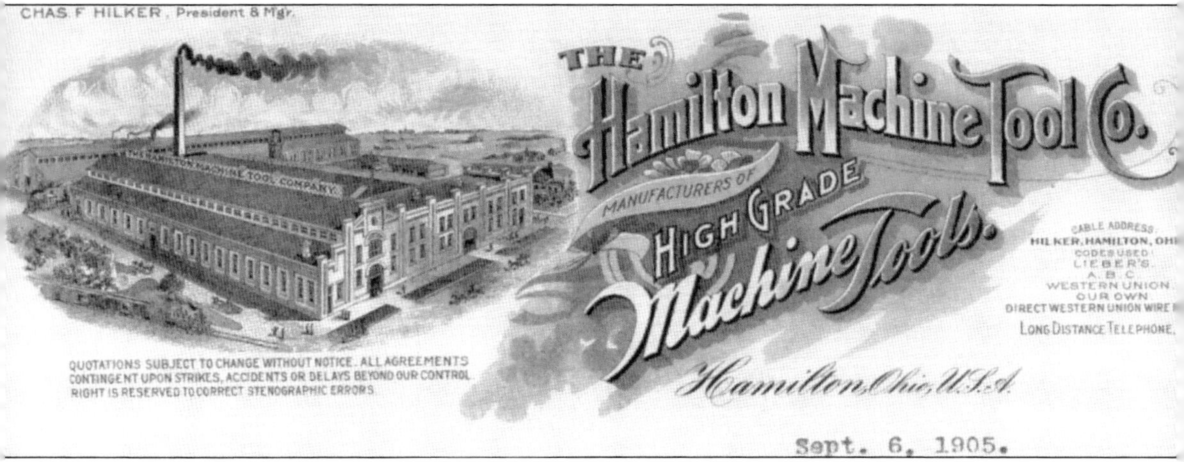

The Hamilton Machine Tool Company manufactured upright drill presses, engine lathes, turret lathes, and similar machine tools. The company was first located along the hydraulic in the building where Charles Hilker's car-seal company operated before the firm built the beginnings of a large plant complex on Symmes Avenue in 1894. That same year, the Shuler and Benninghofen plant also began operations in Lindenwald, and the addition of the two firms stimulated positive growth in the area. In 1905, two years after the company was incorporated, Hilker bought the Belmer Machine Tool Company of Cincinnati and consolidated the two factories at the Hamilton location. The acquisition and the expansion of several product lines led to the enlargement of the Lindenwald plant. An artist's rendering of the factory is featured in the company's letterhead (pictured). Hilker retired from the company in 1923 due to ill health, and the firm went out of business and was deregistered by the Ohio secretary of state in 1926. (Courtesy of Rob Wile.)

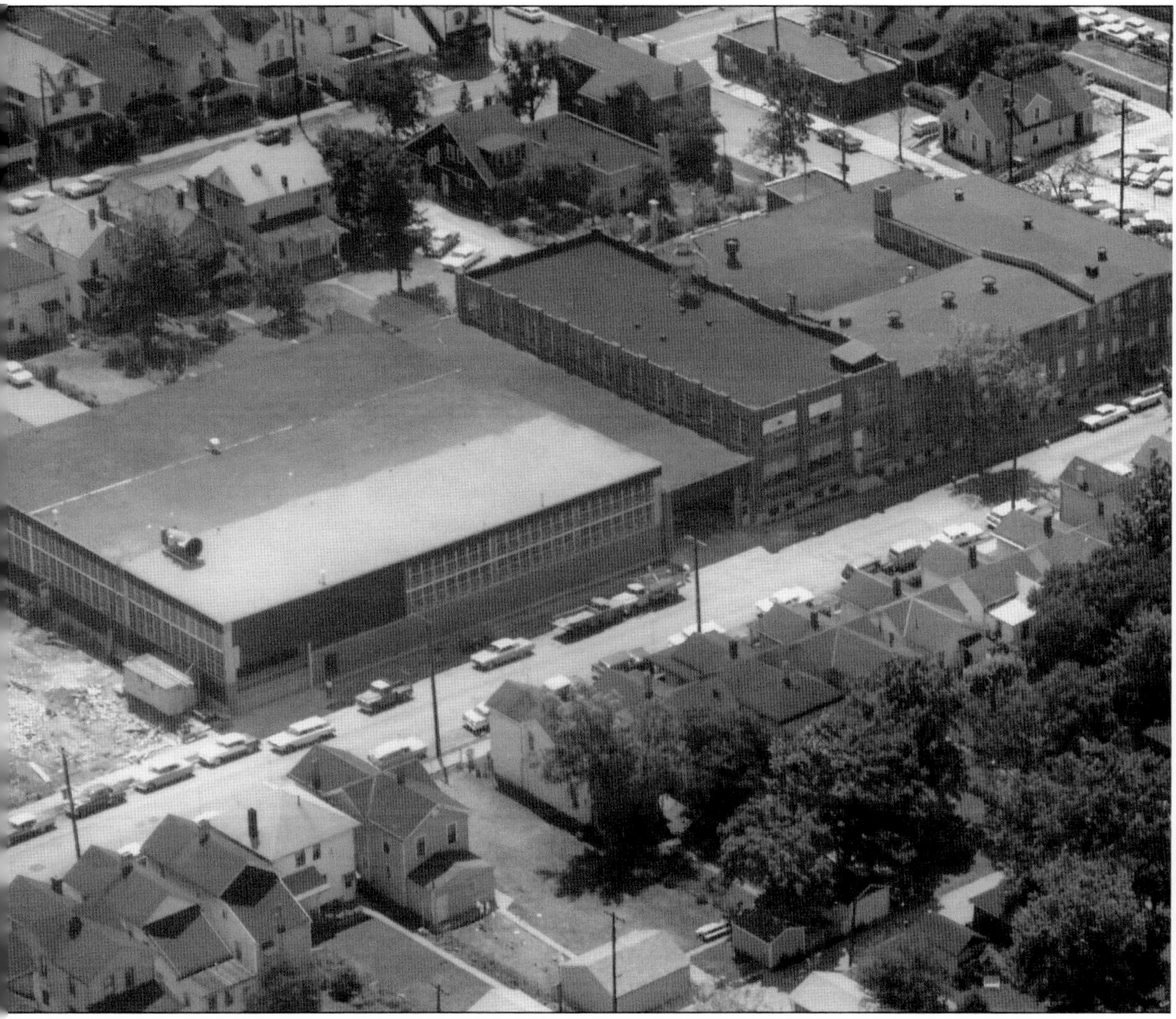

The Hamilton Tool Company began in 1926, after receivers of the bankrupt Hamilton Machine Tool Company agreed to allow the use of a similar name for the new business. The company, organized by Oscar E. Schlichter, designed and manufactured tools, dies, and jigs at its plant on B Street. In 1940, the firm bought the Miami Valley Knitting Mill property at Ninth and Hanover Streets. After World War II, the company shifted to making printing and collating equipment, graphic-arts products, and printing presses, and it had great success. This 1963 aerial photograph of the factory shows several expansions and additions. In 1986, the company merged with the Stevens Graphics Corporation and moved from Hamilton within a few years.

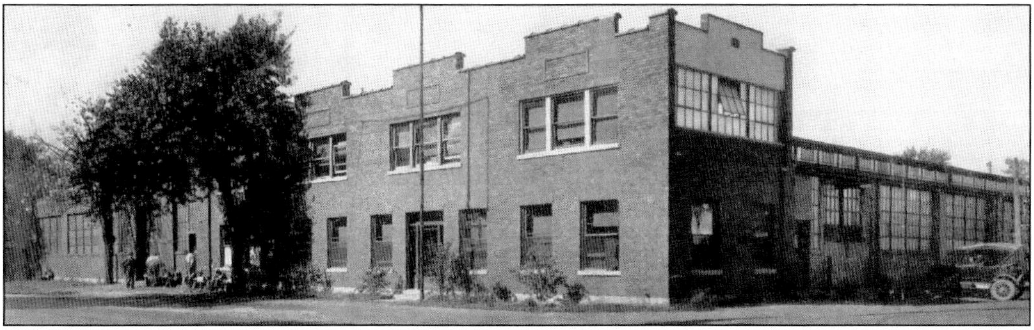

This first factory location of Columbia Machine Tool Company was on Fair Grove Avenue, opposite the Butler County Fairgrounds. The firm had been organized by Frank Yingling in 1910 to manufacture lathes and other machine tools. Yingling also operated the Ceramic Machinery Company, incorporated in 1907, at the same location. In 1937, the company sold its Fair Grove Avenue plant to the Western States Machine Company.

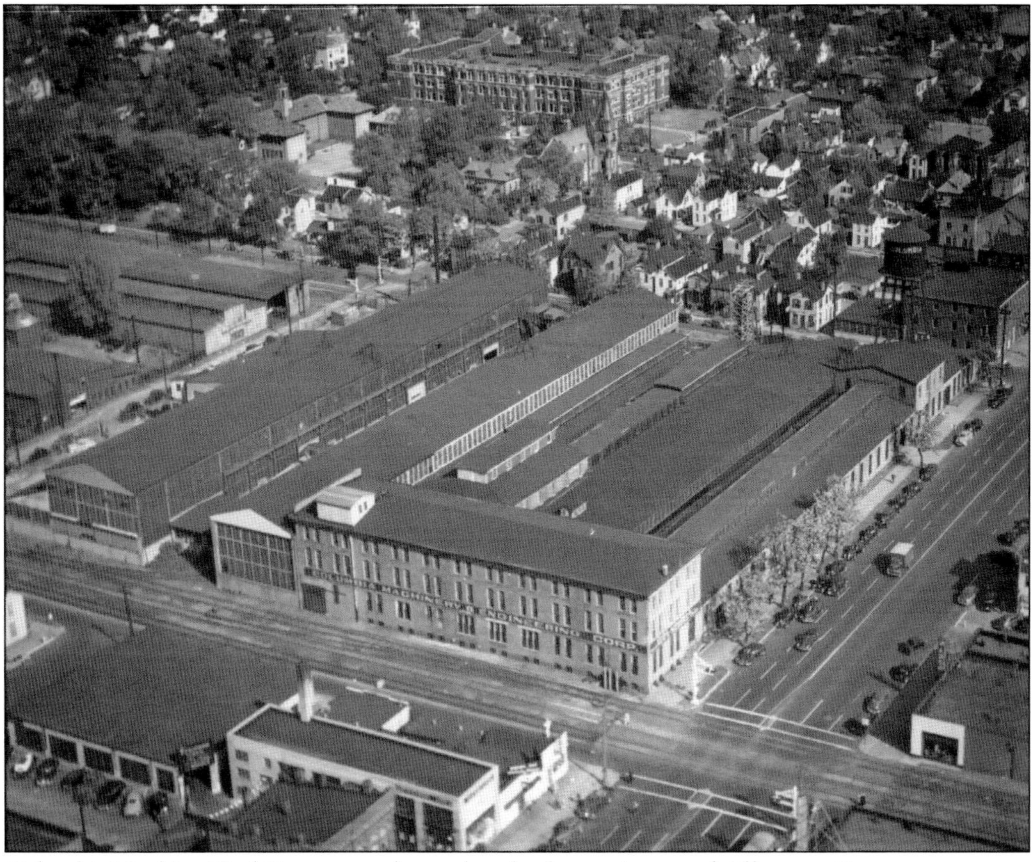

Columbia Machine Tool Company relocated to the former Long and Allstatter Company property on High Street between Fourth and Fifth Streets in 1939. The business was reincorporated as the Columbia Machinery and Engineering Corporation in 1942. The company, which made shears and press brakes, merged with the Lodge and Shipley Company in 1953 and moved all operations to Cincinnati in 1959. The factory, seen here in a 1954 photograph, was razed and made into a parking lot.

Six

Foundries

Many of the items in the lengthy list of products made by Hamilton's factories from the 1850s to the 1950s were produced in one of the foundries that had settled in the city. Most of the early manufacturers had their own foundry operation within their factories to make the metal castings that were parts of a larger product. In some cases, however, a company used a "jobber" foundry for castings. The process was fairly simple: metals such as iron, bronze, aluminum, steel, and brass were heated in extremely hot furnaces until the metal melted into a liquid. The liquid metal was poured into a mold and then removed after the metal cooled and solidified, leaving a casting to be used as needed.

Among the earliest foundries in Hamilton was the Sohn and Rentschler foundry. It was organized in 1875 by Henry Sohn and George Adam Rentschler and had a very successful run of almost 40 years. It ceased doing business in 1913 and turned its contracts over to the Hamilton Foundry and Machine Company, another Rentschler company. Henry P. Deuscher's foundry made castings for farm implements such as planters, churns, and disk harrows, beginning in 1878. The firm adapted to changes in the marketplace and switched to making hot-air furnaces and related products. After a disastrous fire in 1910, the foundry became a "jobber" and only made gray castings for other companies.

Estate Stove Company, Fred J. Meyers Manufacturing Company, and American Frog and Switch made their own respective lines of products. The Estate Stove foundry was one of the largest stove makers in the country, and the Meyers company manufactured virtually everything that could be made from metal wire. American Frog and Switch forged light and medium rails, switches, and other essential railroad products.

Western States Machine Company, which began a relationship with a Hamilton foundry in 1924, moved to nearby Fairfield in 2012. Consequently, the only foundry still operating in Hamilton is the Hamilton Brass and Aluminum Casting Company, formed in 1913.

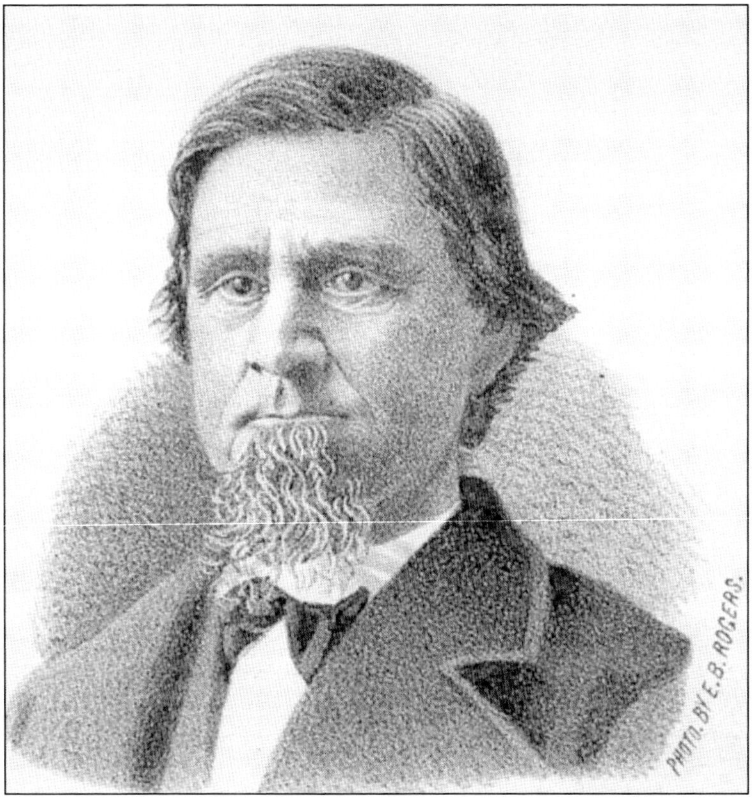

In 1874, German-born Henry P. Deuscher opened a malt house, where he dealt in barley, corn, oats, and malt with Hamilton brewers. After four years, he began operating a foundry in the Variety Iron Works property at Seventh and Hanover Streets. Deuscher, seen here in an 1875 Butler County atlas, incorporated his foundry in 1888. He headed the business until his death in 1903.

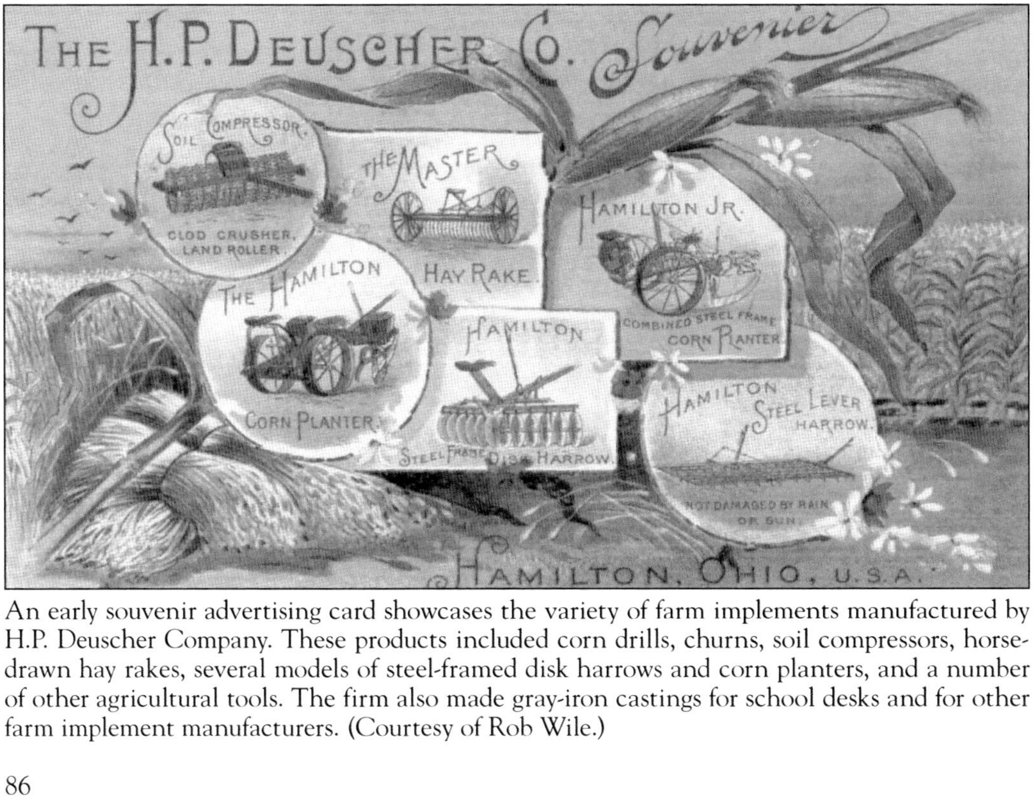

An early souvenir advertising card showcases the variety of farm implements manufactured by H.P. Deuscher Company. These products included corn drills, churns, soil compressors, horse-drawn hay rakes, several models of steel-framed disk harrows and corn planters, and a number of other agricultural tools. The firm also made gray-iron castings for school desks and for other farm implement manufacturers. (Courtesy of Rob Wile.)

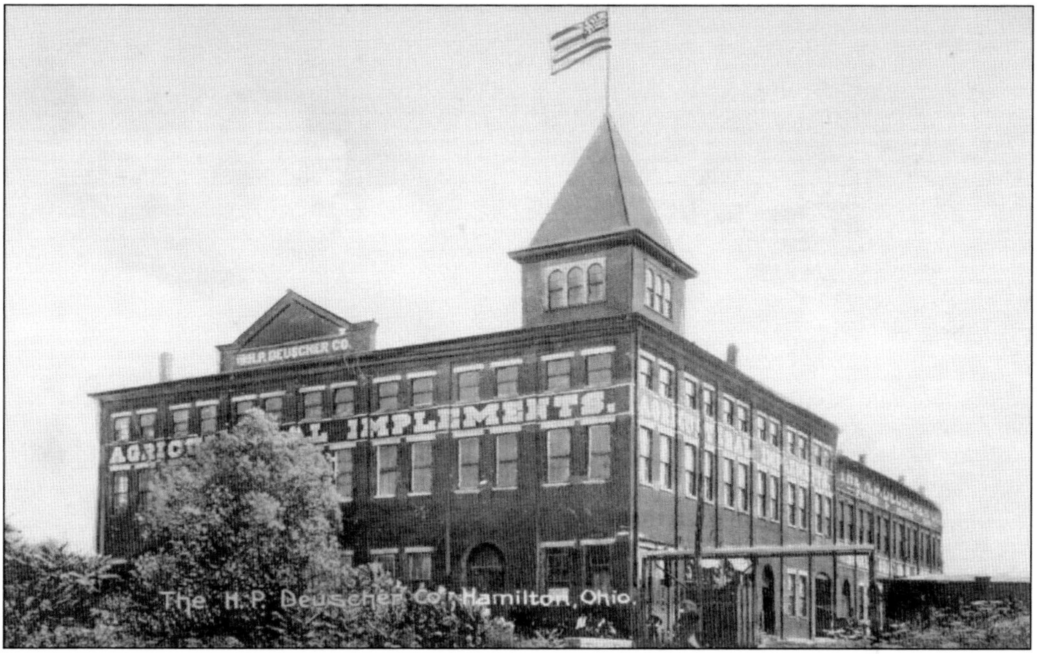

The above postcard, printed before 1910, shows the H.P. Deuscher factory, which had started as the Variety Iron Works. The original foundry had been expanded during the 1890s, when the firm began manufacturing hot-air furnaces and heating and ventilating equipment for a Cincinnati company in 1893. By 1905, Deuscher acquired Cincinnati Heating & Ventilating Company, and much of the firm's production was in this line of products. However, the company continued making horse-drawn farm machinery until 1910. On August 10, 1910, the H.P. Deuscher factory was completely destroyed by fire. The below photograph indicates the degree to which the business was ruined. (Above, courtesy of Rob Wile.)

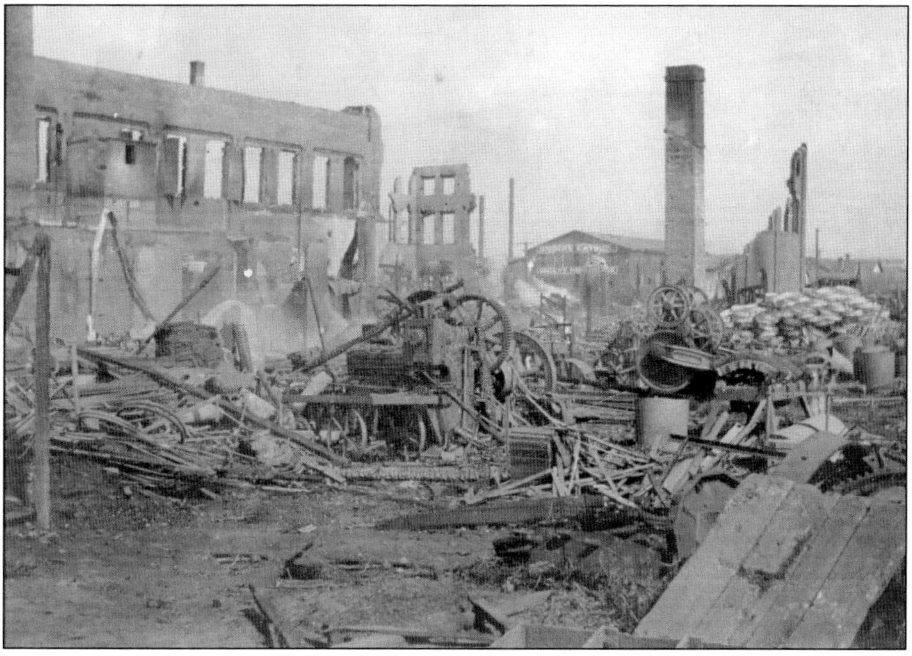

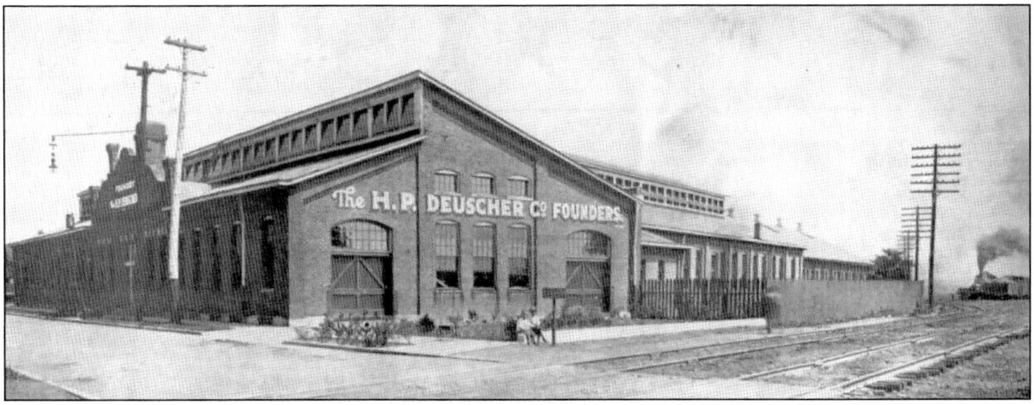

After the 1910 fire, the H.P. Deuscher Company rebuilt its foundry on the same site that it had occupied since 1878. The new plant (pictured) faced Hanover Street, with the foundry running along the tracks of the Cincinnati, Hamilton & Dayton Railroad (right). The company decided to cease manufacturing farm implements entirely and switched to making only gray-metal castings for other companies.

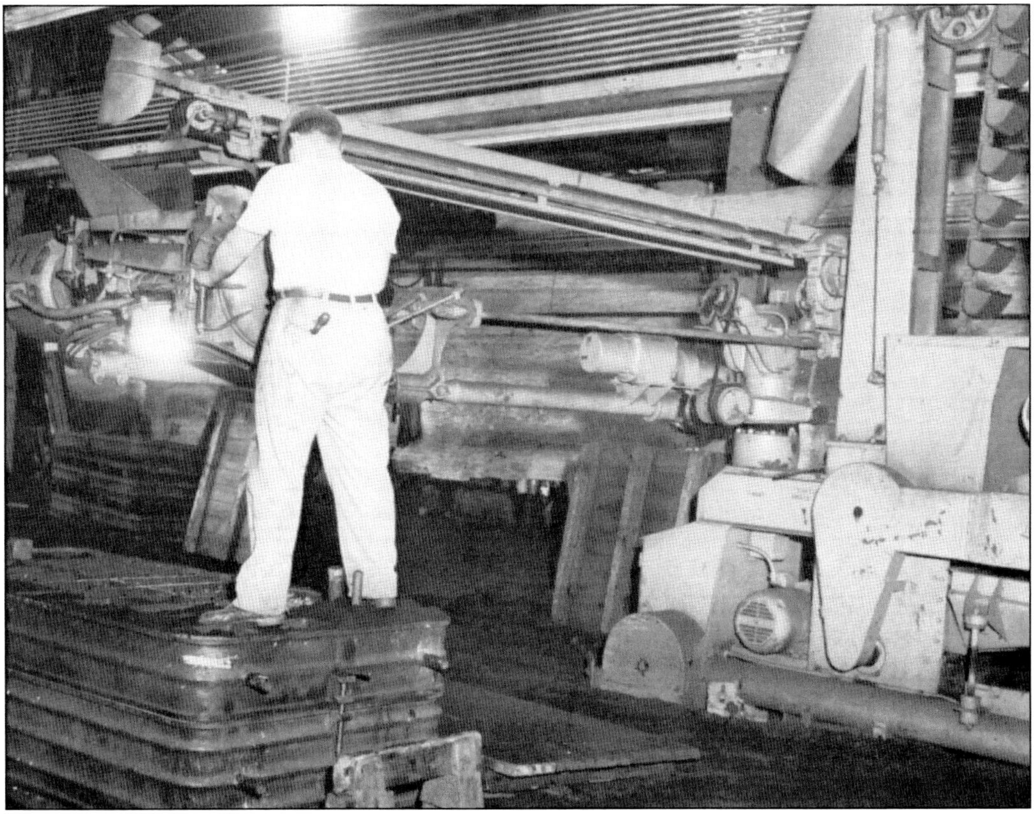

An unidentified employee of H.P. Deuscher works with a mold for a gray-iron casting in the foundry's Motive Sandslinger, used for making larger castings for customers. The machine was installed at the plant in 1954 as part of the company's mechanization-improvement program. In the 1970s and 1980s, the firm made gray-iron castings for pumps, motors, valves, and machine bases for other companies.

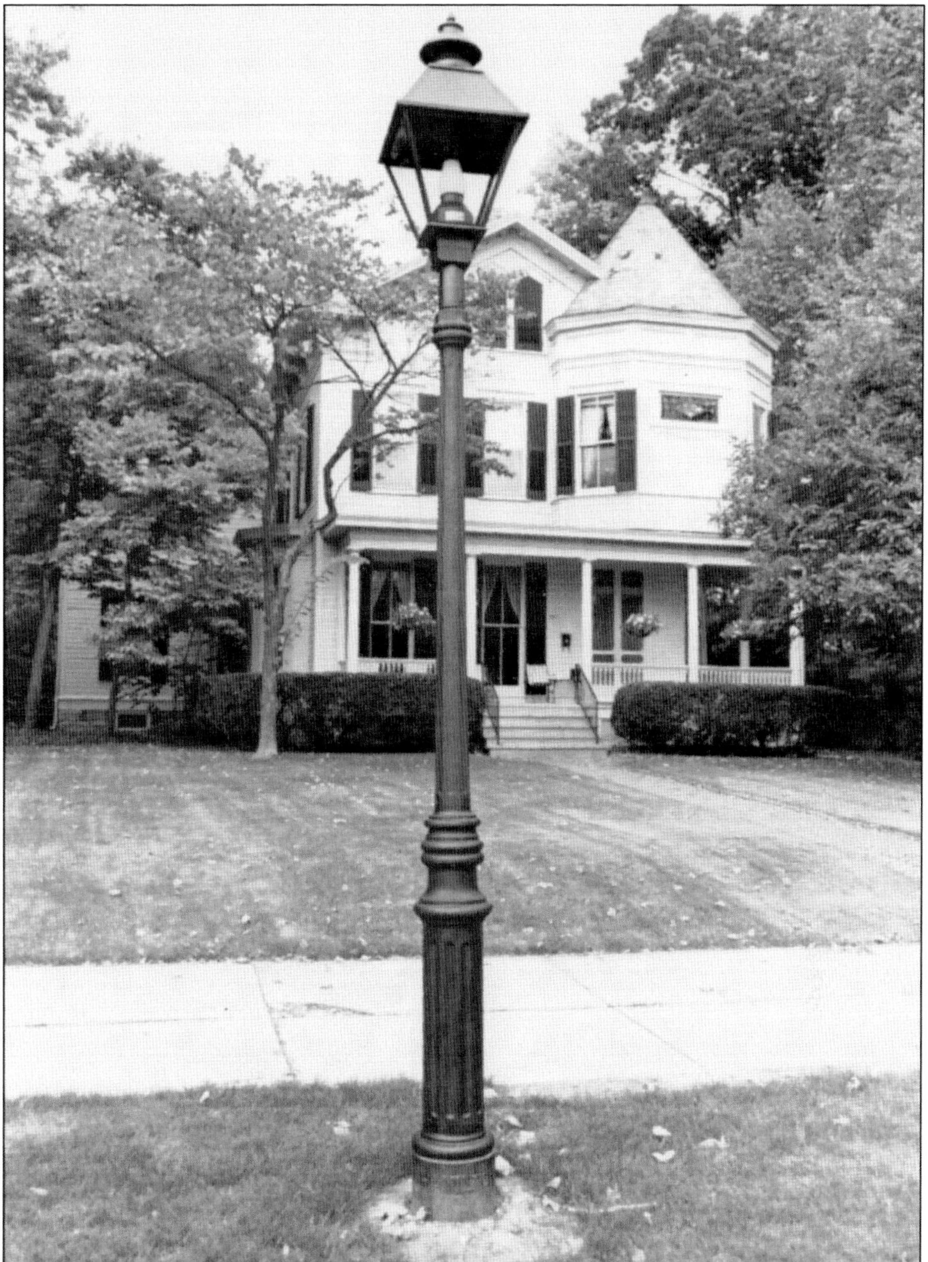

In the 1970s, the H.P. Deuscher Company manufactured decorative lampposts, such as the one seen here. This particular model, the Hamilton Light Post, was installed along streets in Hamilton's German Village district. In 1975, the company installed air pollution control equipment, which removed as much as 99 percent of all particulate matter from the foundry's cupola operation. The company was sold during a leveraged buyout in 1988, ending 110 years of family ownership. The new owners tried to make the foundry profitable, but they were unsuccessful and filed for bankruptcy protection in 1989. All company assets were sold at public auction in September 1990. The H.P. Deuscher factory buildings were razed in August 2000 to make way for a new Butler County jail.

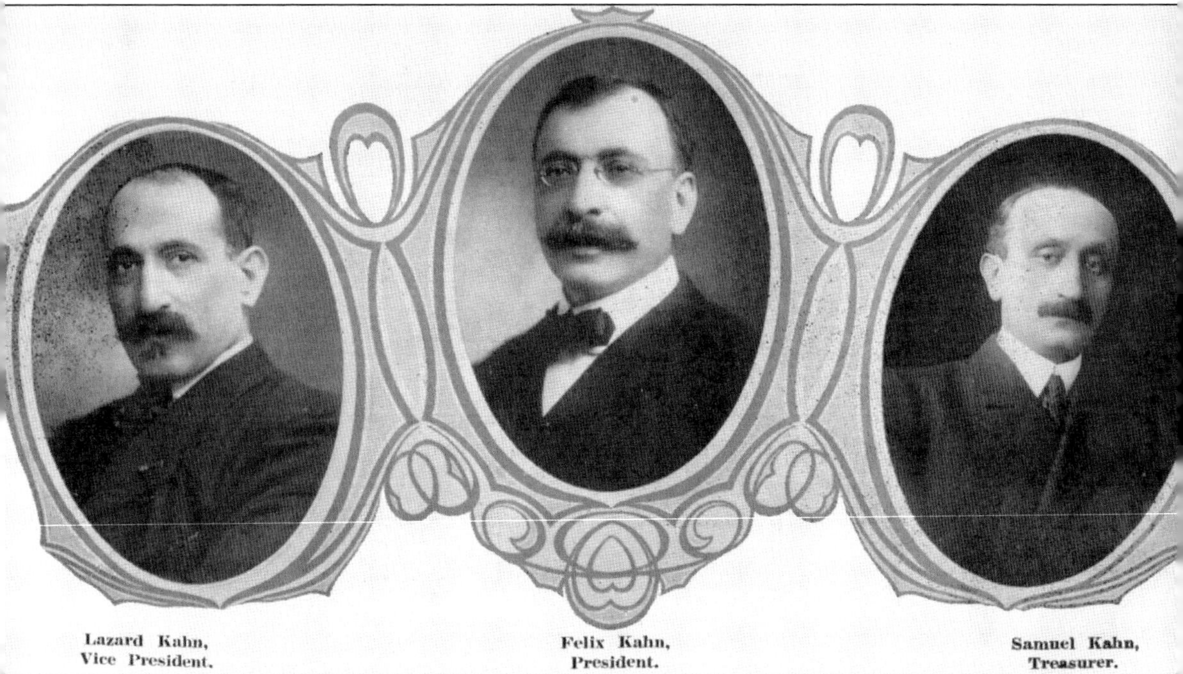

In 1882, Felix, Lazard, and Saul Kahn purchased a stove manufacturer in Hanging Rock, Ohio. The three French immigrants reorganized the company as F. & L. Kahn & Bros. in 1884 and moved the foundry to Hamilton. Saul Kahn died in 1887 and was replaced by another brother, Samuel. The three owners of the company are seen in this composite photograph taken from a 1904 *Daily Republican-News* supplement. Felix (center) handled most of the company's general office and financial matters, and Samuel (right) supervised the outside sales of products. Lazard managed the mechanical aspects of the foundry and its products. Fluent in six languages, he helped develop the firm's international sales. In 1890, Lazard joined with Moses Mosler and Oakey Parrish to form the Hamilton Improvement Syndicate for the purpose of attracting other businesses to the city. The group purchased land on the east side of Hamilton and persuaded several manufacturers, including Mosler Safe and Lock Company, Hamilton Foundry and Machine Company, and the Fred J. Meyers Manufacturing Company, to locate in the area.

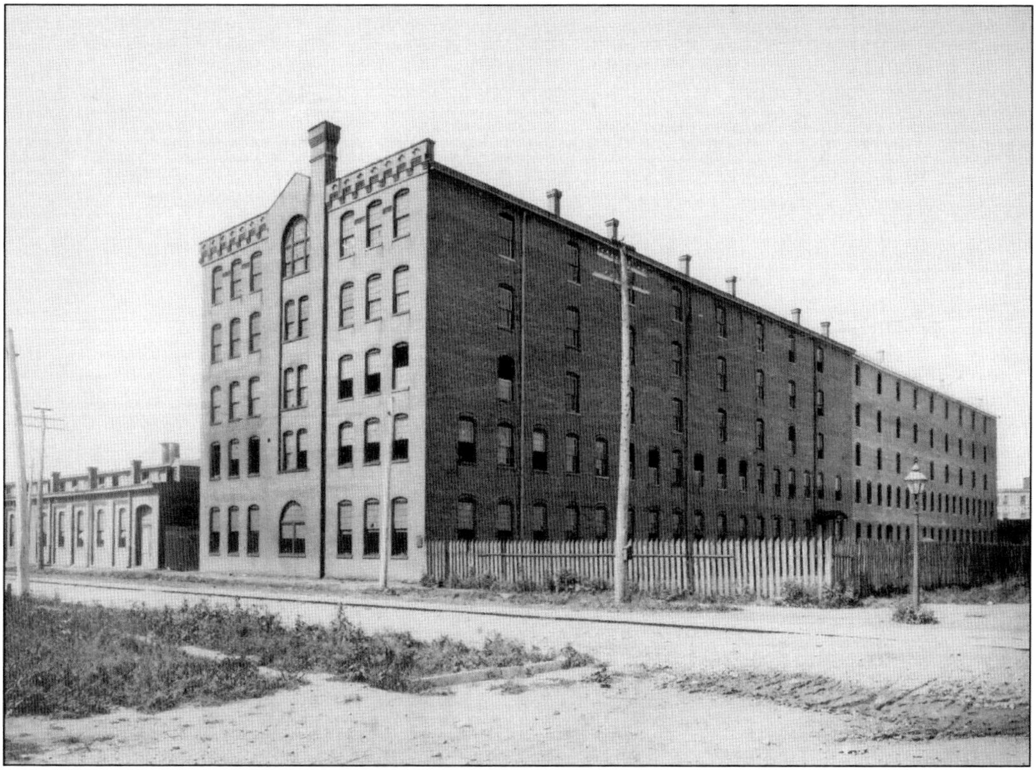

This five-story building was the factory that F. & L. Kahn & Bros. occupied less than five years after the foundry began production in 1885. There had been expansions in 1889 and 1890 at the left of the main building, and an addition at the rear. Additional expansions would be made as the company's product lines were developed. For example, the foundry began making gas stoves in 1888.

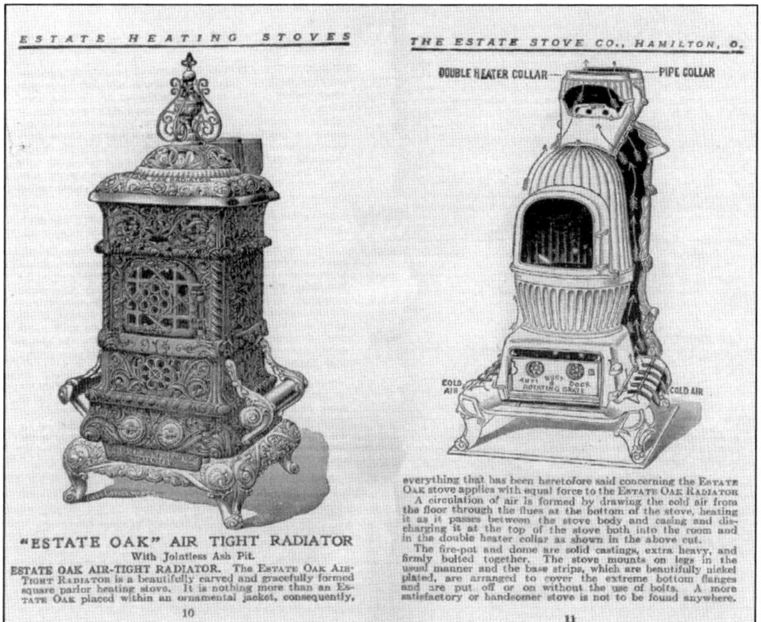
ESTATE HEATING STOVES

THE ESTATE STOVE CO., HAMILTON, O.

"ESTATE OAK" AIR TIGHT RADIATOR

With Jointless Ash Pit.

ESTATE OAK AIR-TIGHT RADIATOR. The Estate Oak Air-Tight Radiator is a beautifully carved and gracefully formed square parlor heating stove. It is nothing more than an Estate Oak placed within an ornamental jacket, consequently, everything that has been heretofore said concerning the Estate Oak stove applies with equal force to the Estate Oak Radiator.

A circulation of air is formed by drawing the cold air from the floor through the flues at the bottom of the stove, heating it as it passes between the stove body and casing and discharging it at the top of the stove both into the room and in the double heater collar as shown in the above cut.

The fire-pot and dome are solid castings, extra heavy, and firmly bolted together. The stove mounts on legs in the usual manner and the base strips, which are beautifully nickel plated, are arranged to cover the extreme bottom flanges and are put off or on without the use of bolts. A more satisfactory or handsomer stove is not to be found anywhere.

10

11

An early catalog page shows the type of cast-iron coal- or wood-burning radiator furnaces made by F. & L. Kahn & Bros. The "Estate Oak" brand (left) was one of the most popular brands the company made. The model seen here dates from 1892, the year Lazard Kahn received a patent for the jointless ash pit feature. The "Estate" brand name had been used by the company as early as 1880. (Courtesy of Rob Wile.)

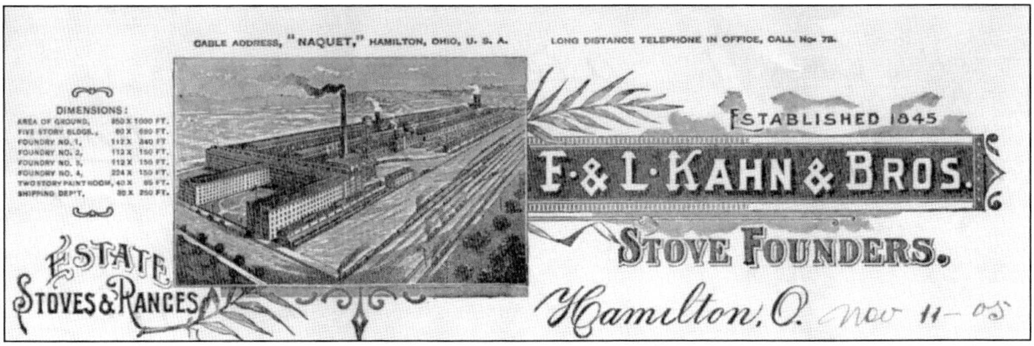

The letterhead seen here lists the sizes of the components that made up the F. & L. Kahn & Bros. factory. The company had three five-story buildings, four foundries, a two-story paint shop, shipping building, an engine house, coal shed, and coke shed. The letterhead is dated November 11, 1905, less than two months before the company was incorporated as the Estate Stove Company on December 28, 1905. (Courtesy of Rob Wile.)

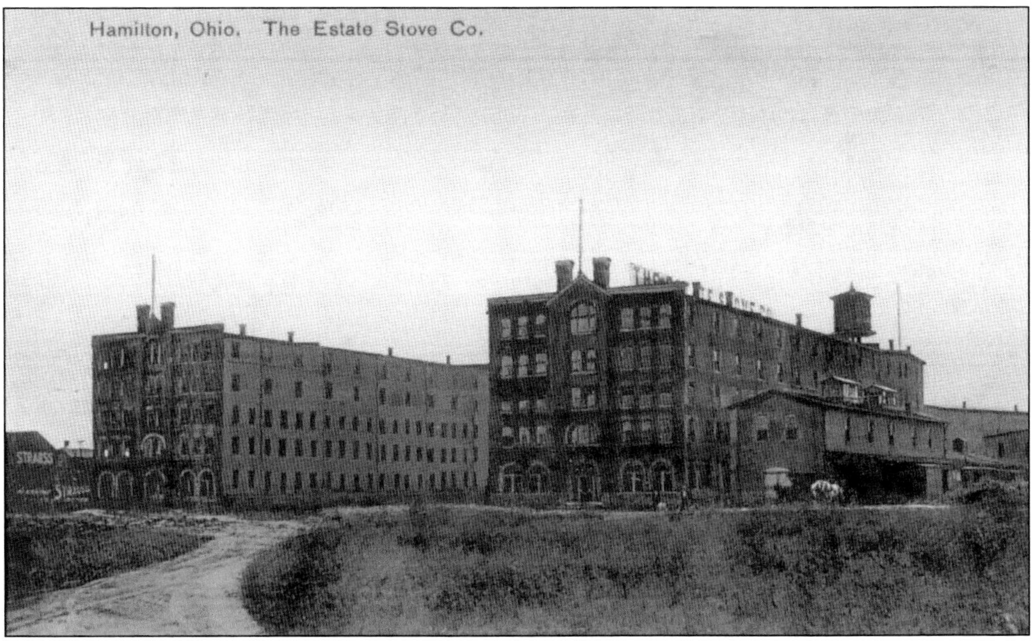

This c. 1905 postcard offers a better view of the Estate Stove Company plant. Shown here are two of the five-story buildings and the paint shop (right). The buildings are accessed by an unpaved road, which was used by the horse-drawn carriage in front of the two-story paint shop. The company's increased production and new product lines, such as the Estate Heatrola electric heaters in 1921, led to expanding the size of the factory.

The Estate Stove Company, like nearly all Hamilton businesses, took part in the three-day Preparedness Day observation on August 2, 1940. The event included programs featuring patriotic displays and speeches by community leaders and military personnel. The company's float for the parade, seen here leaving the factory, carried Heatrola and Estate appliances, as well as the message that whatever helped business promoted national defense.

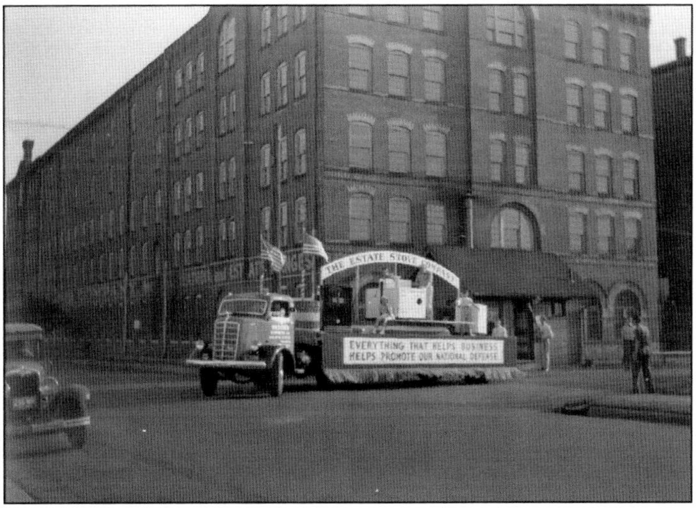

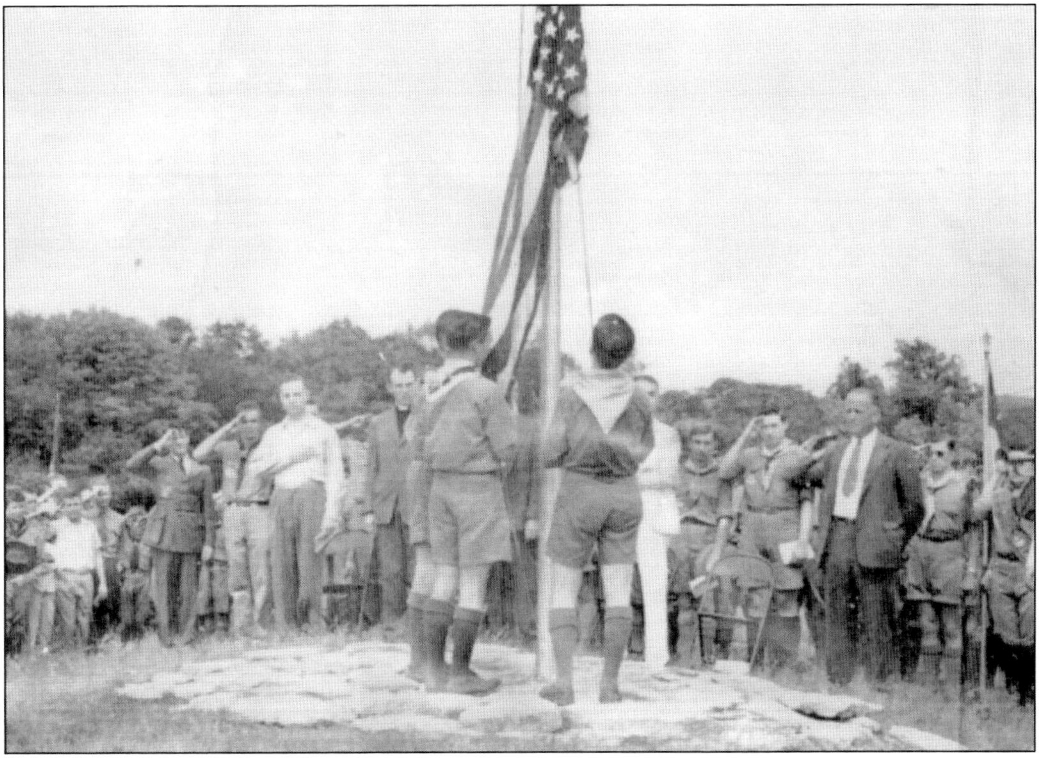

This June 21, 1942, photograph was taken during a flag-raising ceremony dedicating a new Boy Scout Camp. The camp was named Camp Myron Kahn in honor of Estate Stove's secretary and son of founder Samuel Kahn, who had died in 1937. The man in the coat and tie at the right is Lucian L. Kahn, Estate Stove's treasurer and son of founder Lazard Kahn. The Pleasant Vineyard Ministries acquired the scout camp for nondenominational programs in 1981.

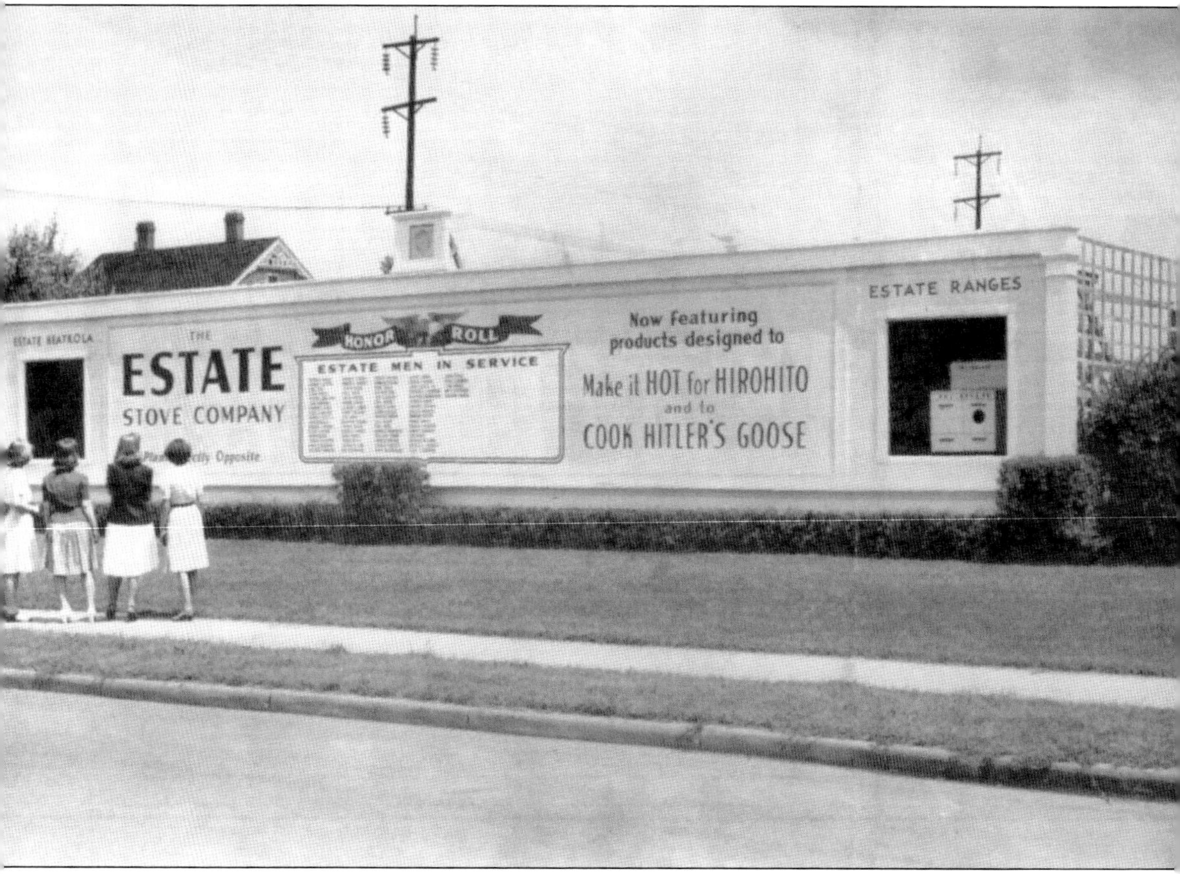

This billboard was located on East Avenue, opposite the Estate Stove Company's factory. It listed the names of the company's employees serving in the armed forces during World War II. At the time of this photograph, 69 names were listed. During the war, Estate Stove manufactured parts for tanks and airplanes, ammunition shells, anti-tank mine casings, and a variety of other military supplies. After the war, in 1945, the Noma Electric Corporation bought the company. Another sale took place in 1952, when the Radio Corporation of America bought the firm, forming the RCA Estate Appliance Corporation. In 1955, another merger placed the company under the control of the Whirlpool-Seeger Corporation. Whirlpool announced that production would cease at Hamilton's Estate Stove plant by June 1961.

Henry Sohn (right), George A. Rentschler (below), and John Balle established a foundry they named the Ohio Iron Works on Vine Street at the hydraulic in 1875. Balle sold his interest to his partners in 1876. For more than 35 years, Sohn and Rentschler built and maintained a highly successful and profitable company. Both men were also owners of the Hooven, Owens, Rentschler Company, Phoenix Caster Company, and an icehouse. They had other business interests as well. In 1891, Rentschler formed the Hamilton Foundry and Machine Company in a 100-foot-by-200-foot building on Lincoln Avenue in East Hamilton. Sohn and Rentschler's Ohio Iron Works foundry was swept away by the 1913 Great Miami River flood. The partners decided not to rebuild and instead turned over all Ohio Iron Works contracts to the Hamilton Foundry and Machine Company.

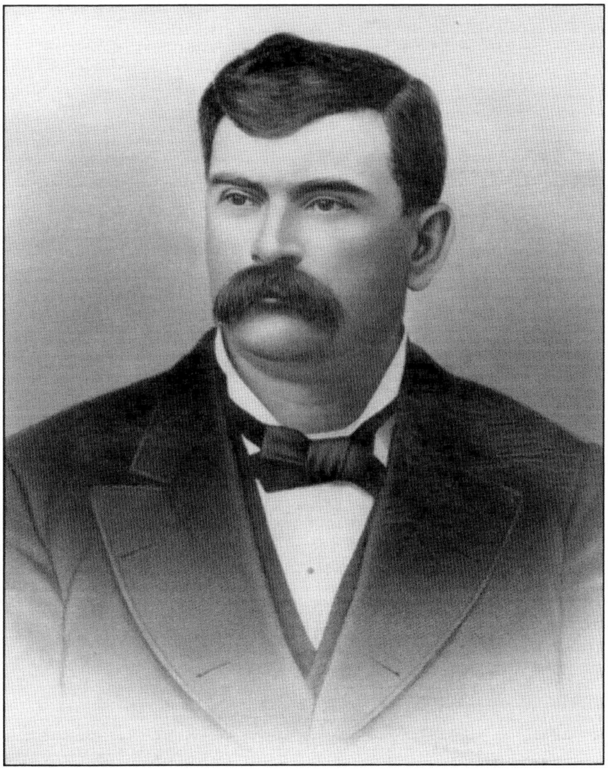

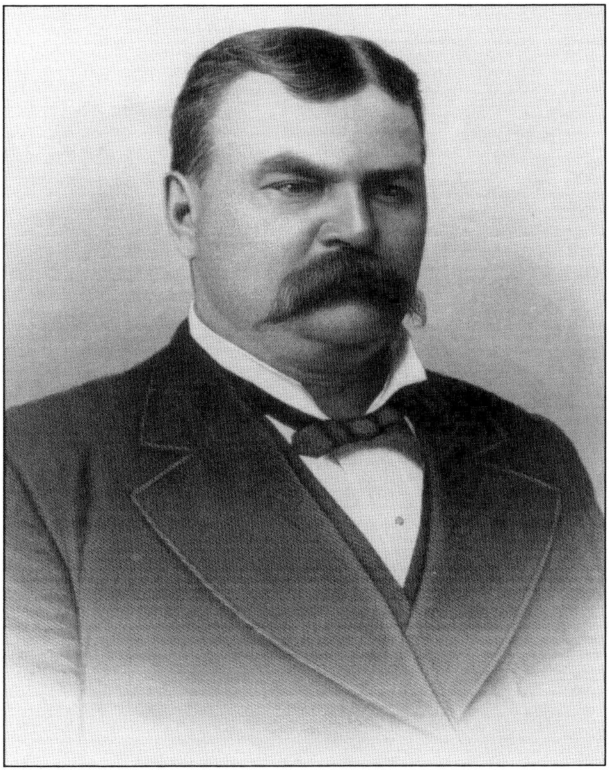

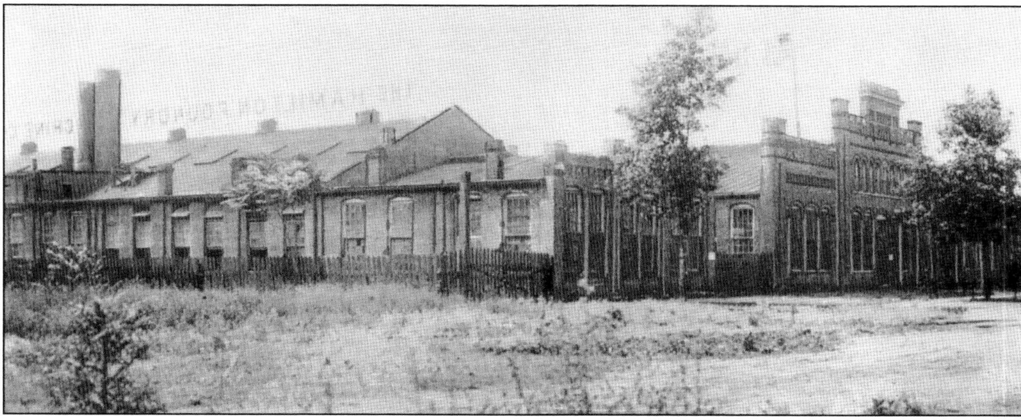

Before 1906, the Hamilton Foundry and Machine Company occupied these buildings on an unpaved Lincoln Avenue south of the Mosler Safe Company. The building at right was the original 200-foot-by-100-foot foundry, with a 100-foot-by-100-foot addition for pattern storage, storeroom, and office. The building to the left was an 80-foot-by-55-foot wing addition for a woodworking shop, grinding room, emery room, and shipping.

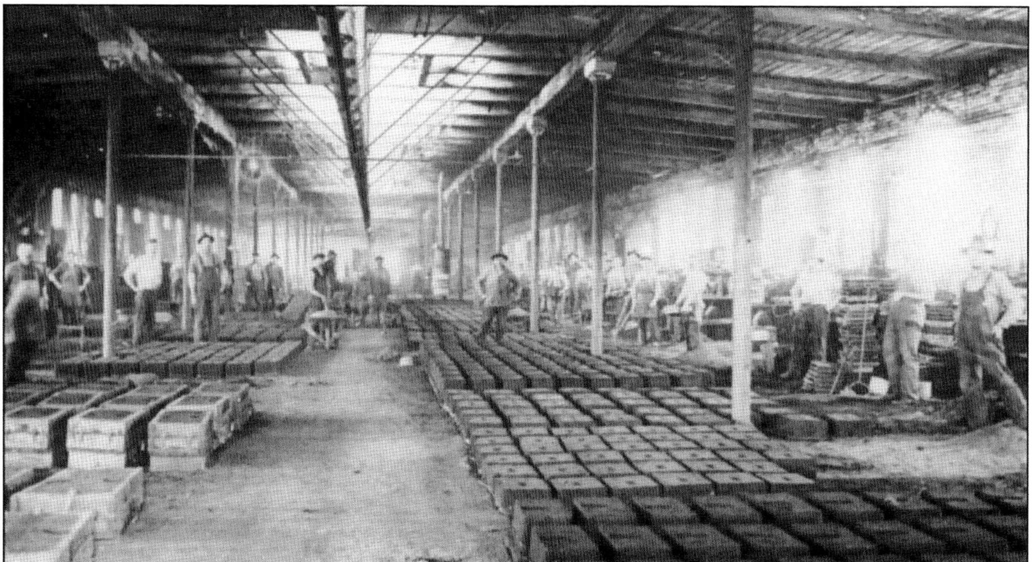

This photograph, dating to the World War I era, was taken inside the Hamilton Foundry and Machine Company's No. 1 foundry. The neatly arranged molds on the floor were the forms in which workers poured hot metal to make gray-iron casting for customers. The only person identified in the photograph is Joe Fager, the man wearing a sweater standing in the center.

A group of unidentified women employees of the Hamilton Foundry and Machine Company stand outside the facility, holding an Army-Navy "E" Award flag. The firm received its first "E" award on September 23, 1943. This photograph dates from around April 1944, because the flag has an additional white star the firm had earned for maintaining outstanding performance for six additional months. The company would receive a total of four white stars by the end of the war.

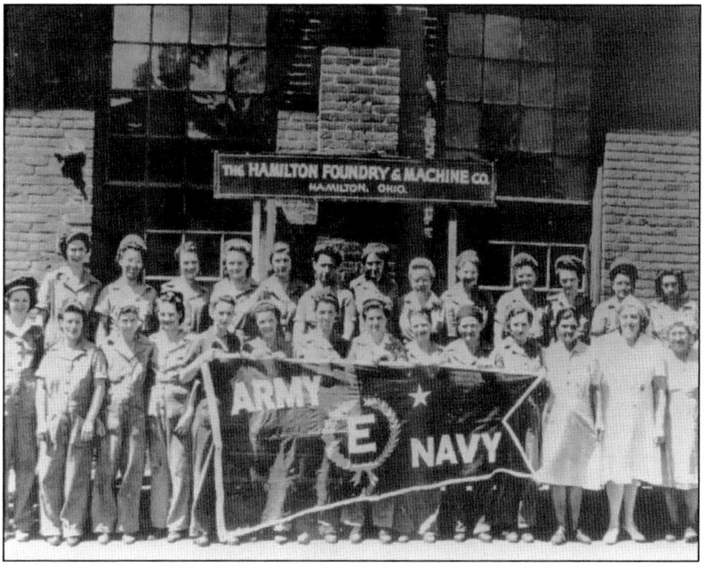

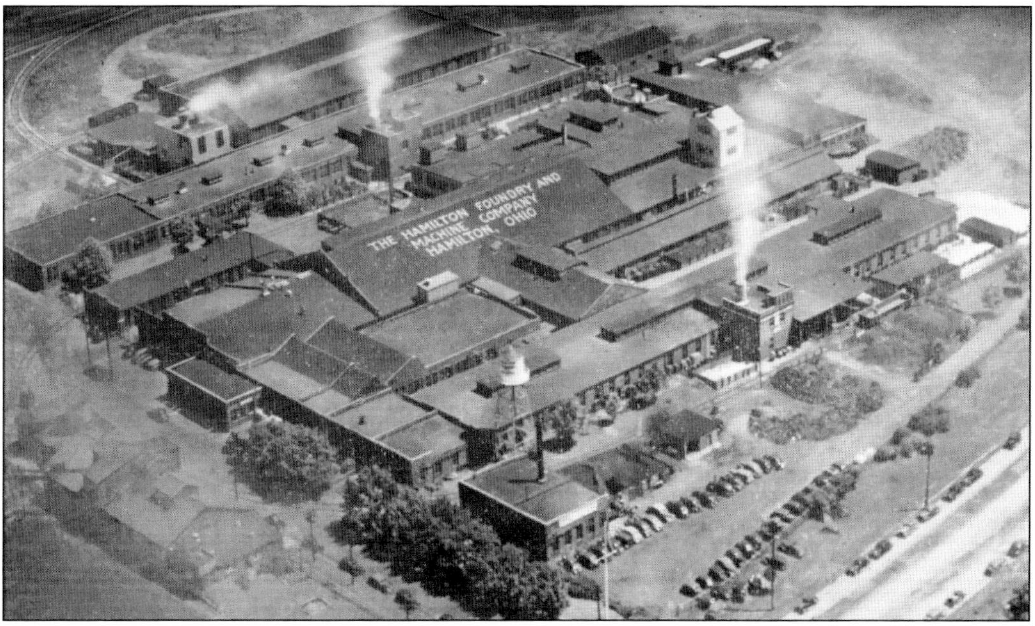

This photograph of the Hamilton Foundry and Machine Company plant appeared in the firm's 1953 new employee handbook. In the 1960s, the factory manufactured products for companies in a 15-state area. The plant was remodeled and expanded, and hundreds of rose bushes were planted along the foundry's 900-foot-long frontage on Erie Highway. Production at the plant ceased in 1984, and the factory was demolished in 1987.

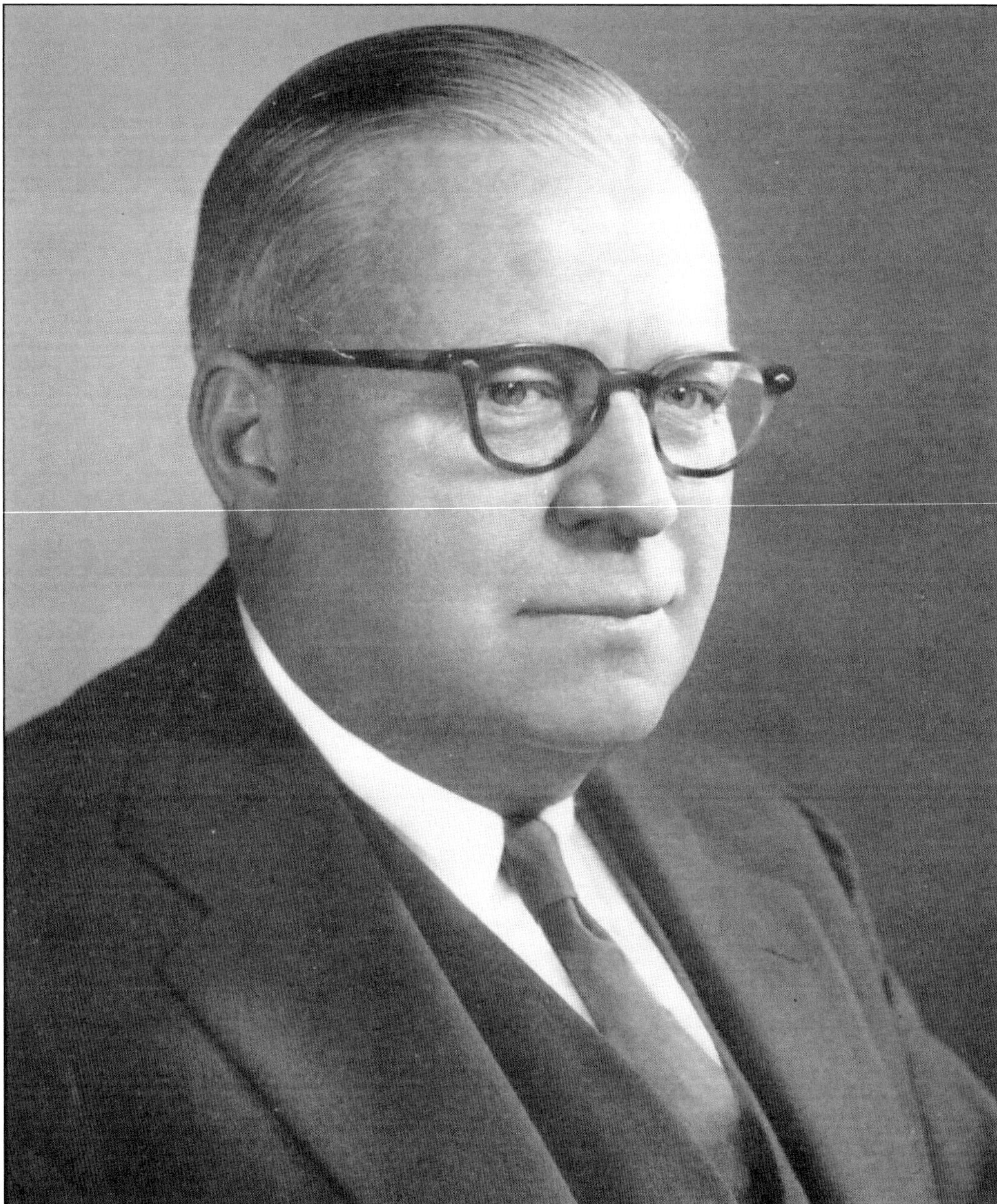

Peter Rentschler, grandson of Hamilton Foundry and Machine Company founder George A. Rentschler, joined the company in 1920. He became vice president in 1923 and president in 1927, serving in that office until his death in 1976. Some of his community-service activity included his work with Reuben Robertson of Champion Coated Paper Company to establish the Hamilton Safety Council in 1942. Rentschler served on the Hamilton Board of Education from 1940 to 1947 and chaired the 1965 fundraising committee that raised more than $1.5 million to help establish the Hamilton campus of Miami University. The campus library building was named in his honor. During his administration of the Hamilton Foundry and Machine Company, the firm's name was changed to Hamilton Foundry, Inc., in 1959 and then to Hamilton Allied Corporation in 1969.

Fred J. Meyers, seen right in a 1911 *Daily Republican-News* cartoon, began his company in Covington, Kentucky, in 1865. He incorporated his firm as the Fred J. Meyers Manufacturing Company in 1883. Meyers had a thriving business when his five-story factory was destroyed by fire in 1893. He decided to move his company to Hamilton and built a three-story brick plant on Bender Avenue. The artist's rendering below is a depiction of the factory Meyer built in 1893 and the one-story brick shop used by his firm. The company made an extremely diverse range of products out of metal and wire, including bird cages, kitchen utensils, fancy grillwork, wrought-iron fencing, lawn furniture, iron and steel jail cells, bicycles, bank and office fixtures, elevator cars and enclosures, railroad conductor punches, and streetcar fenders.

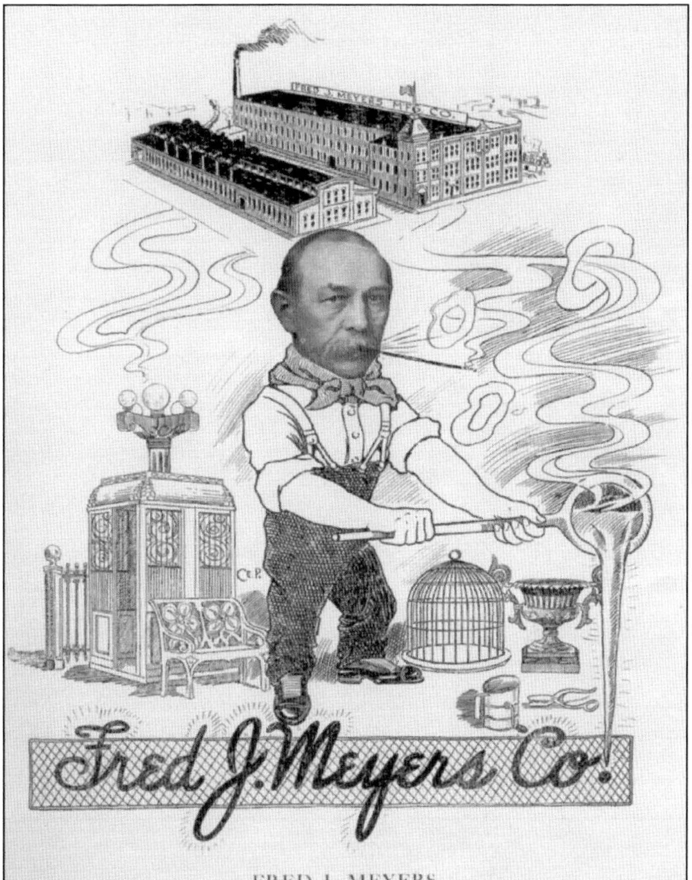

FRED J. MEYERS

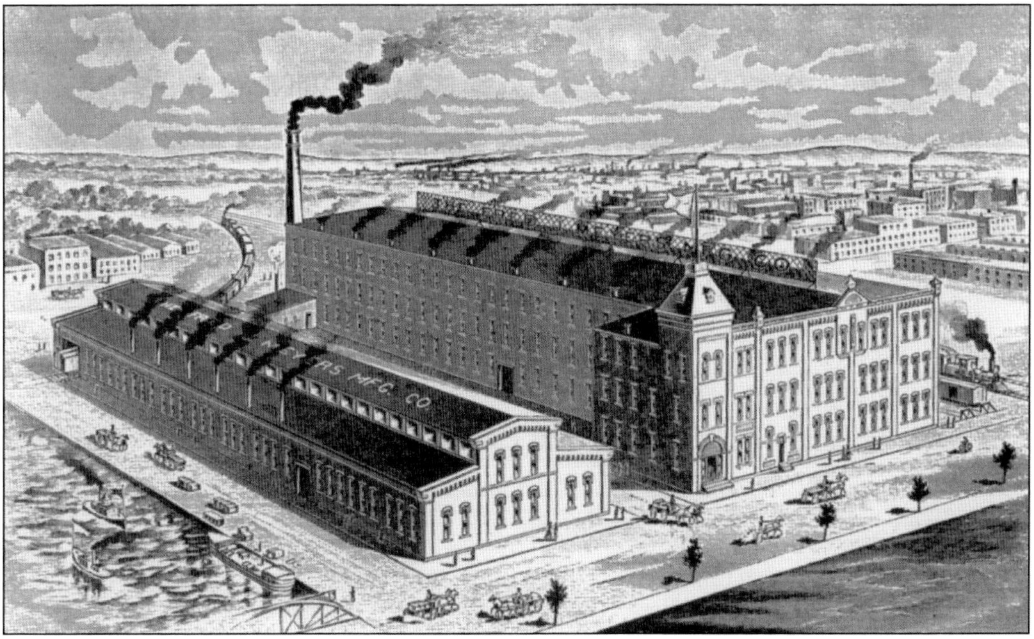

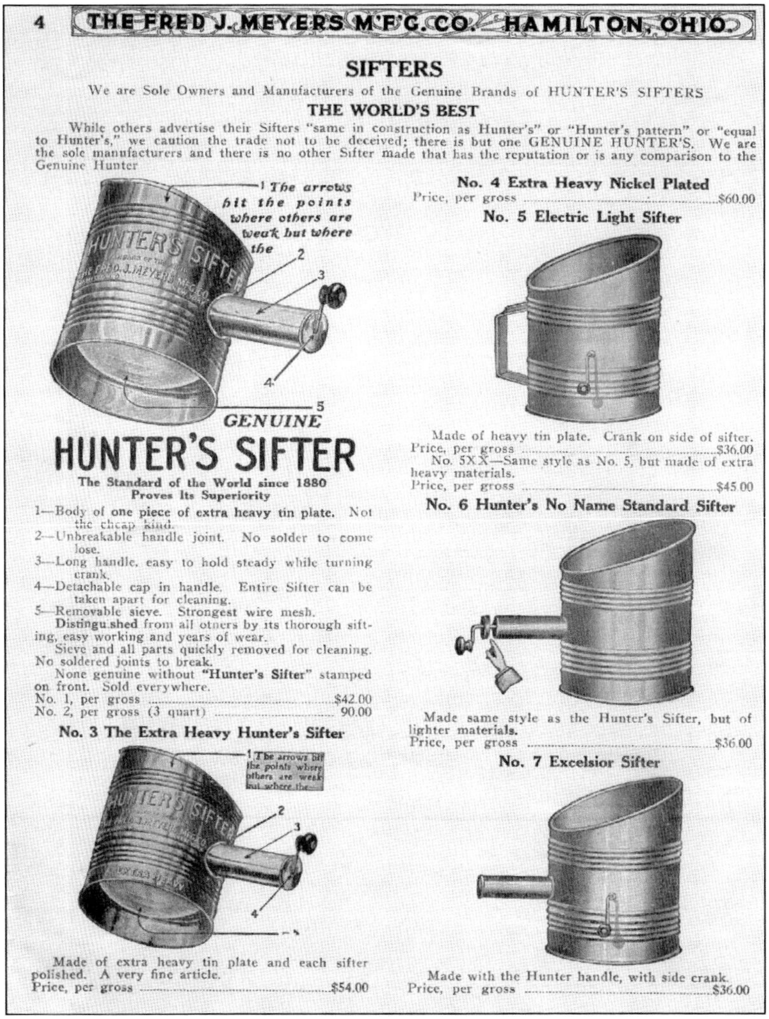

4 THE FRED J. MEYERS M'F'G. CO. HAMILTON, OHIO.

SIFTERS

We are Sole Owners and Manufacturers of the Genuine Brands of HUNTER'S SIFTERS

THE WORLD'S BEST

While others advertise their Sifters "same in construction as Hunter's" or "Hunter's pattern" or "equal to Hunter's," we caution the trade not to be deceived; there is but one GENUINE HUNTER'S. We are the sole manufacturers and there is no other Sifter made that has the reputation or is any comparison to the Genuine Hunter

GENUINE

HUNTER'S SIFTER

The Standard of the World since 1880
Proves Its Superiority

1—Body of one piece of extra heavy tin plate. Not the cheap kind.
2—Unbreakable handle joint. No solder to come lose.
3—Long handle, easy to hold steady while turning crank.
4—Detachable cap in handle. Entire Sifter can be taken apart for cleaning.
5—Removable sieve. Strongest wire mesh.

Distingu shed from all others by its thorough sifting, easy working and years of wear.

Sieve and all parts quickly removed for cleaning. No soldered joints to break.

None genuine without **"Hunter's Sifter"** stamped on front. Sold everywhere.

No. 1, per gross $42.00
No. 2, per gross (3 quart) 90.00

No. 3 The Extra Heavy Hunter's Sifter

Made of extra heavy tin plate and each sifter polished. A very fine article.
Price, per gross $54.00

No. 4 Extra Heavy Nickel Plated
Price, per gross $60.00

No. 5 Electric Light Sifter

Made of heavy tin plate. Crank on side of sifter.
Price, per gross $36.00
No. 5XX—Same style as No. 5, but made of extra heavy materials.
Price, per gross $45.00

No. 6 Hunter's No Name Standard Sifter

Made same style as the Hunter's Sifter, but of lighter materials.
Price, per gross $36.00

No. 7 Excelsior Sifter

Made with the Hunter handle, with side crank.
Price, per gross $36.00

The Hunter's flour sifter was one of the Fred J. Meyers Manufacturing Company's most popular products. Each of the sifters seen on this page from a catalog for wholesalers was built from a single piece of extra heavy tin plate, had a long handle, and could be used as a mixer, scoop, measurer, rice washer, and strainer. The wholesale cost for a gross of the Hunter's Sifter at the upper left was $42, or about 30¢ each.

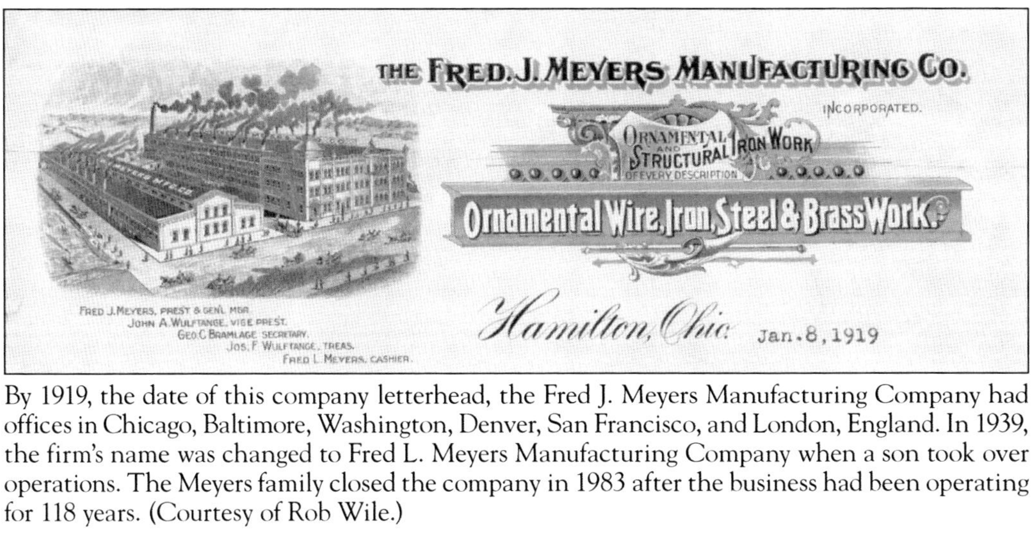

By 1919, the date of this company letterhead, the Fred J. Meyers Manufacturing Company had offices in Chicago, Baltimore, Washington, Denver, San Francisco, and London, England. In 1939, the firm's name was changed to Fred L. Meyers Manufacturing Company when a son took over operations. The Meyers family closed the company in 1983 after the business had been operating for 118 years. (Courtesy of Rob Wile.)

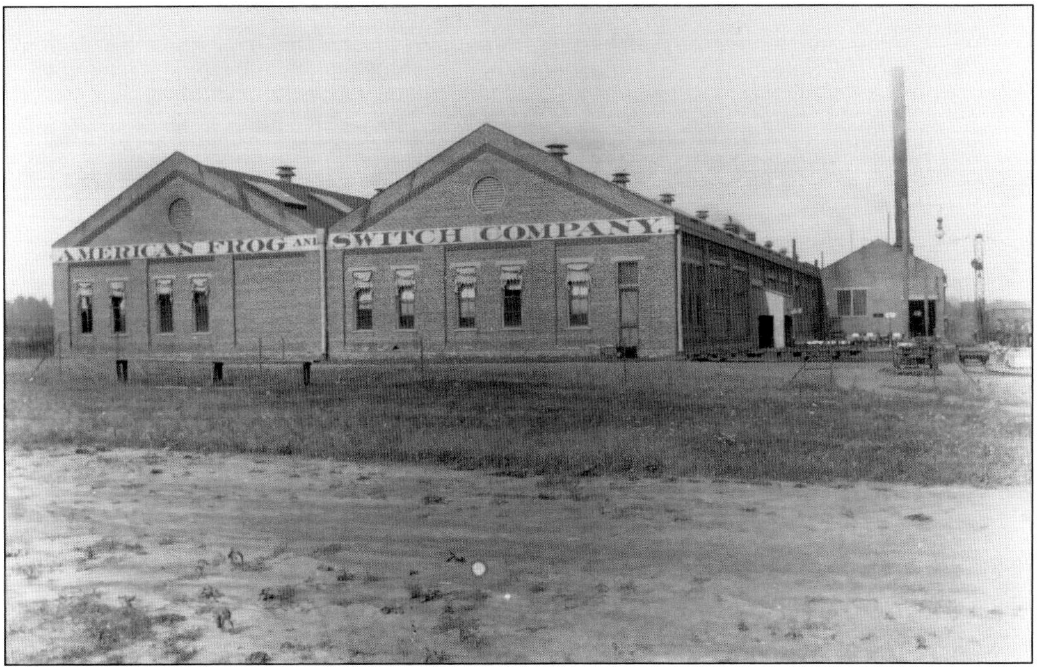

This building, located on West Main Street at Haldimand Avenue, was the factory for American Frog and Switch Company, makers of railroad-track turnouts, switches, and special rail supplies. The company was formed in 1901 by a group of Cincinnati and Hamilton businessmen, including John C. Hooven, president of Hooven, Owens, Rentschler, and Charles E. Heiser, president of the Second National Bank. (Courtesy of Dick Haid.)

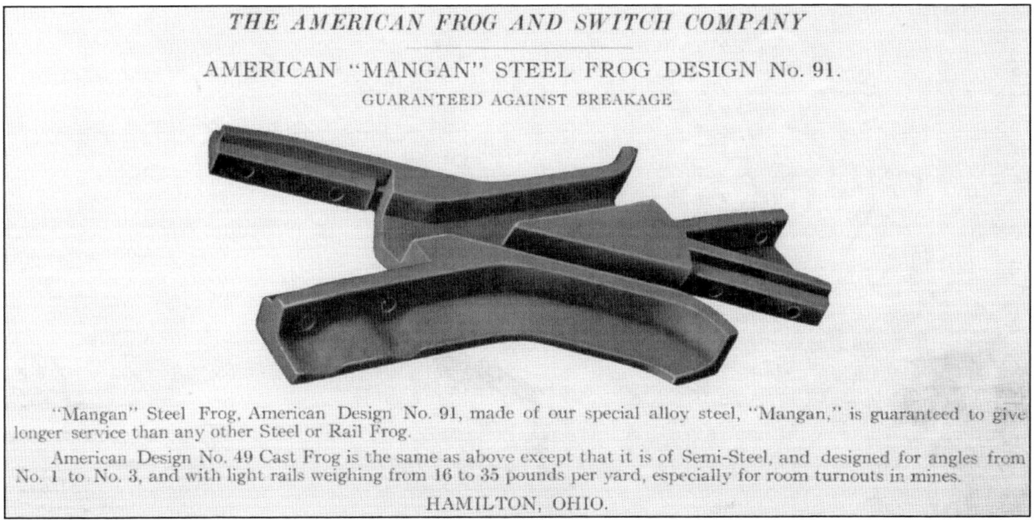

THE AMERICAN FROG AND SWITCH COMPANY

AMERICAN "MANGAN" STEEL FROG DESIGN No. 91.

GUARANTEED AGAINST BREAKAGE

"Mangan" Steel Frog, American Design No. 91, made of our special alloy steel, "Mangan," is guaranteed to give longer service than any other Steel or Rail Frog.

American Design No. 49 Cast Frog is the same as above except that it is of Semi-Steel, and designed for angles from No. 1 to No. 3, and with light rails weighing from 16 to 35 pounds per yard, especially for room turnouts in mines.

HAMILTON, OHIO.

An example of a steel frog made by American Frog and Switch is seen in this page from a company catalog. This piece of rail, placed where two rail lines intersected, enabled railcars to travel across the intersection or change from one line to the other line. This particular frog was made for light-rail applications, but the company made heavier frogs for use in US Navy yards, military installations, and railroads throughout the world. (Courtesy of Dick Haid.)

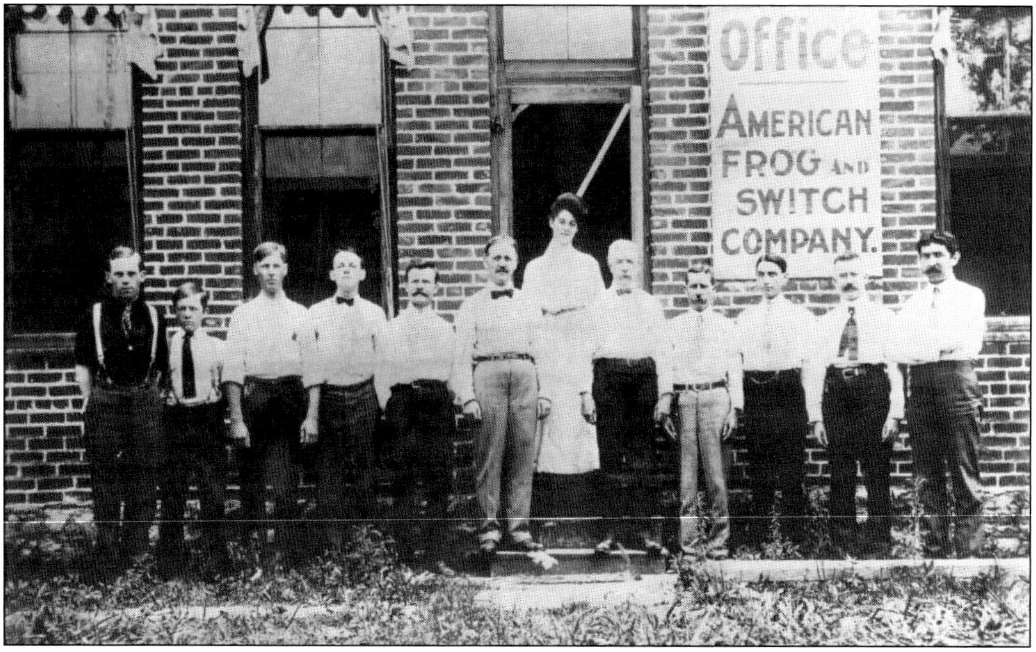

Employees and officers of American Frog and Switch Company stand in front of their office building in 1910. They are, from left to right, James Haskell, unidentified, Elmer Byerline, Paul Conner, John W. Rabbe, Eugene S. Griffis (treasurer), Clara Schweinfest, L.F. Phipps (president), M.E. Dempsey (secretary), George Remp, E.W. Harden (superintendent), and William H. Rabbe (chief engineer). (Courtesy of Dick Haid.)

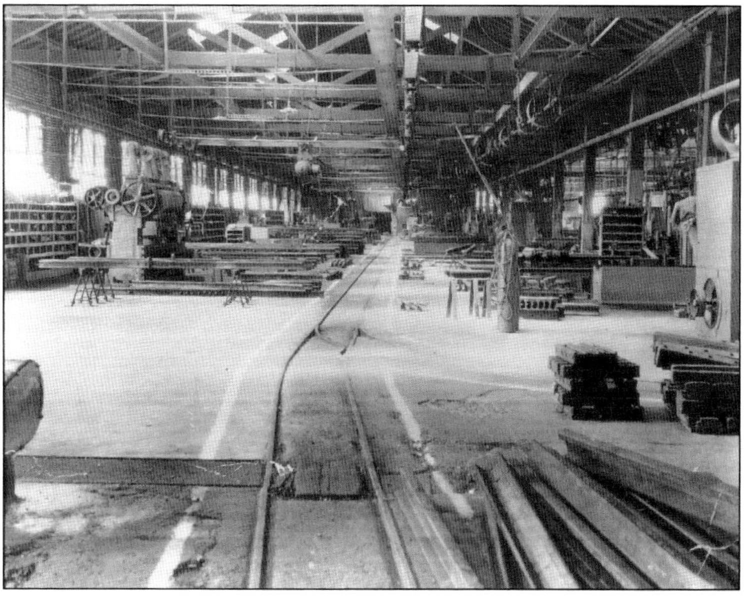

The south bay of American Frog and Switch Company is seen during relatively prosperous times in this 1928 photograph. The firm was reorganized under supervision of a bankruptcy court in 1934, when it was sold to the Taylor-Wharton Iron and Steel Company of New Jersey. Production ceased at the plant in May 1949, and the factory was razed. The Ohio secretary of state dissolved the company on October 22, 1949.

The American Foundry and Machine Company began making small and medium-sized gray-iron castings in May 1900. After five years, the firm was incorporated on July 6, 1905. The foundry's plant, depicted on the catalog cover shown right, was located between Fairview and Forest Avenues, near Zimmerman Avenue in the Lindenwald area. The company shifted to manufacturing gas ranges, cooking stoves, hot plates, and a variety of heating equipment for the consumer market. Shown left in a page in one of the company's catalog for consumers is one of the foundry's Cupid line of gas ranges. Product sales were not sufficient to keep the company solvent. The business was dissolved by the Ohio secretary of state on October 14, 1929.

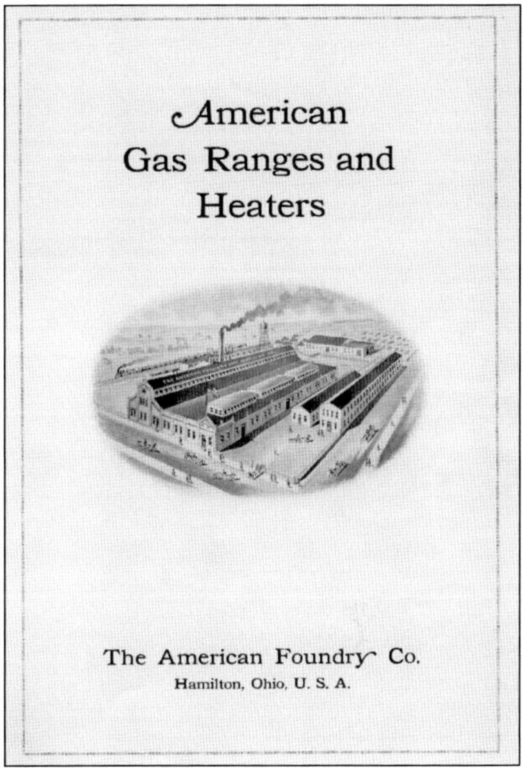

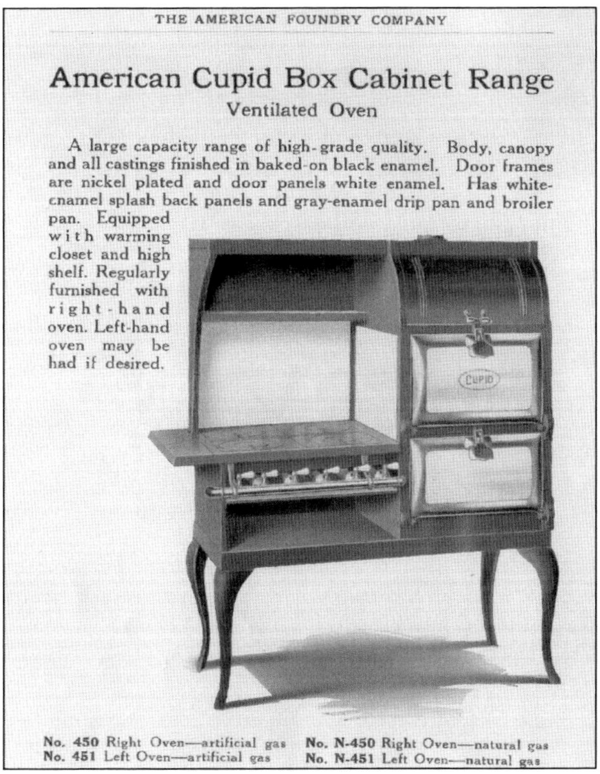
THE AMERICAN FOUNDRY COMPANY

American Cupid Box Cabinet Range

Ventilated Oven

A large capacity range of high-grade quality. Body, canopy and all castings finished in baked-on black enamel. Door frames are nickel plated and door panels white enamel. Has white-enamel splash back panels and gray-enamel drip pan and broiler pan. Equipped with warming closet and high shelf. Regularly furnished with right-hand oven. Left-hand oven may be had if desired.

No. 450 Right Oven—artificial gas
No. 451 Left Oven—artificial gas
No. N-450 Right Oven—natural gas
No. N-451 Left Oven—natural gas

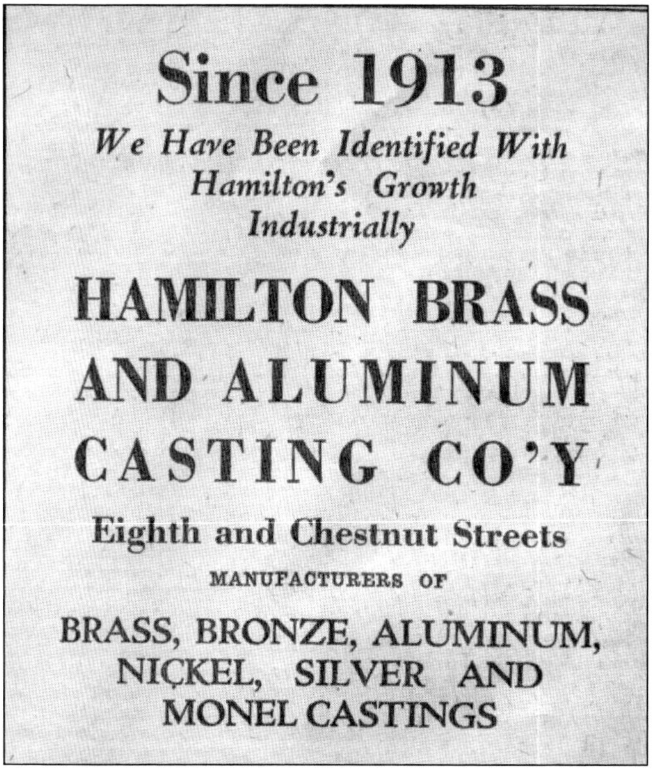
Since 1913

We Have Been Identified With Hamilton's Growth Industrially

HAMILTON BRASS AND ALUMINUM CASTING CO'Y

Eighth and Chestnut Streets

MANUFACTURERS OF

BRASS, BRONZE, ALUMINUM, NICKEL, SILVER AND MONEL CASTINGS

Charles E. Koehler began Hamilton Brass and Aluminum Casting Company in 1913. It was incorporated on June 26, 1918. This advertisement appeared in the *Hamilton Journal News* in 1936. Charles T. Koehler, the founder's son, formed Centri-Cast Corporation in 1955 and the Miami-Cast Foundry in Miamisburg in 1993. Hamilton Brass and Aluminum Castings Company recovered from a destructive fire in 1976, rebuilt its factory, and, today, is the second-oldest manufacturing firm still operating in Hamilton.

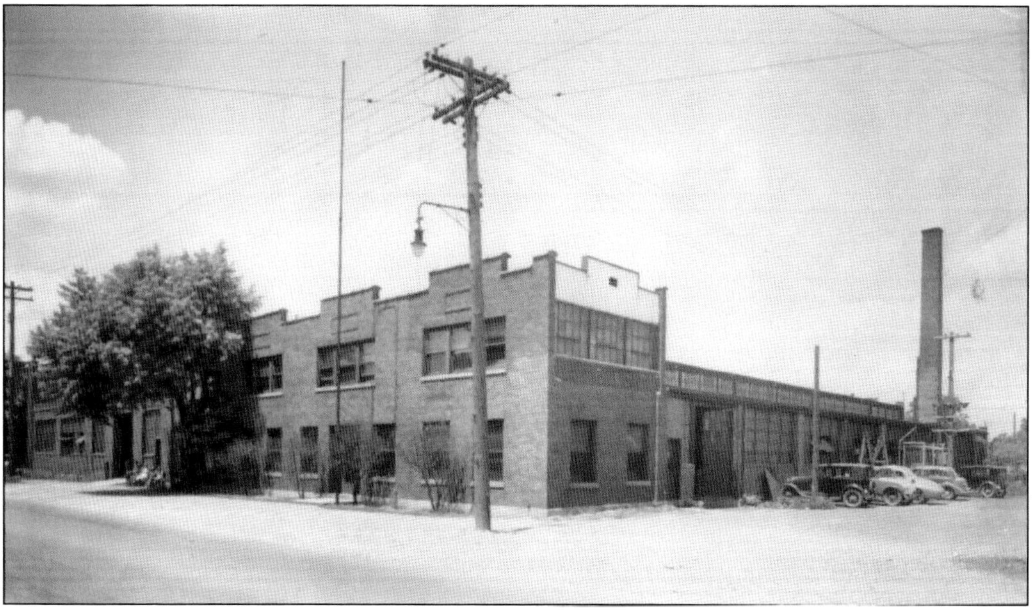

In 1917, Western States Machine Company began making centrifuge equipment in Salt Lake City, Utah, for the sugar industry. The firm entered an agreement in 1924 whereby Columbia Machine Tool Company of Hamilton manufactured some of its centrifuge equipment. In 1936, Western States bought the machine tool company's building (pictured) and moved some of its operations to Hamilton. The firm built a new manufacturing facility and office in Fairfield in 2012.

Seven

Other Manufacturers

The variety and diversity of the products manufactured in Hamilton helped set the city apart from other manufacturing towns. In many categories of product lines, such as papermaking, safes, machine tools, and engines, several factories were engaged in making the same items. In some industries, some of the companies were among the largest in the world. This was especially the case in papermaking, manufacturing safes, making machine tools, and forging stoves. In many smaller shops and factories, skilled and trained workers in virtually all industrial professions contributed to making Hamilton a great manufacturing city.

Shuler and Benninghofen's Miami Woolen Mills was a prominent maker of blankets and clothing and had a large percentage of the market for papermaking felts. The Miami Valley Knitting Mills was a leader in the manufacture of underwear, especially the one-piece union suit. The Leshner Corporation evolved into working with fabrics, including sheets and towels.

Examples of firms that made machinery for specialized purposes include the Triumph Laundry Machinery Company and the Hamilton Autographic Register Company. The former made a variety of machines to clean and iron sheets and clothing. The latter company manufactured devices that made multiple copies of bills of lading and receipts. Nearly every retail outlet in the United States used an autographic register product.

Beverages made by Hamilton companies included beers brewed by Martin Mason Brewing Company and Cincinnati Brewing Company and fruit-flavored soft drinks, bottled waters, ciders, and vinegars made by J. Pabst and Sons.

Milling grain into flour was a dominant area of activity in the early years of industrial Hamilton. Semler Milling and Carr Milling were two of the last flour mills that operated in town.

Other products made in Hamilton included buckets, mops, and brooms produced by the Eagle Woodenware Company; ornamental and decorative tiles made by the Hamilton Tile Works; music stands by Krauth and Benninghofen; coolers made by the Hamilton Skotch Company; casters made by Phoenix Casters and Hamilton Caster & Manufacturing; tin cans and other packaging by the American Can Company; and water closets made by the Sanitary Manufacturing Company.

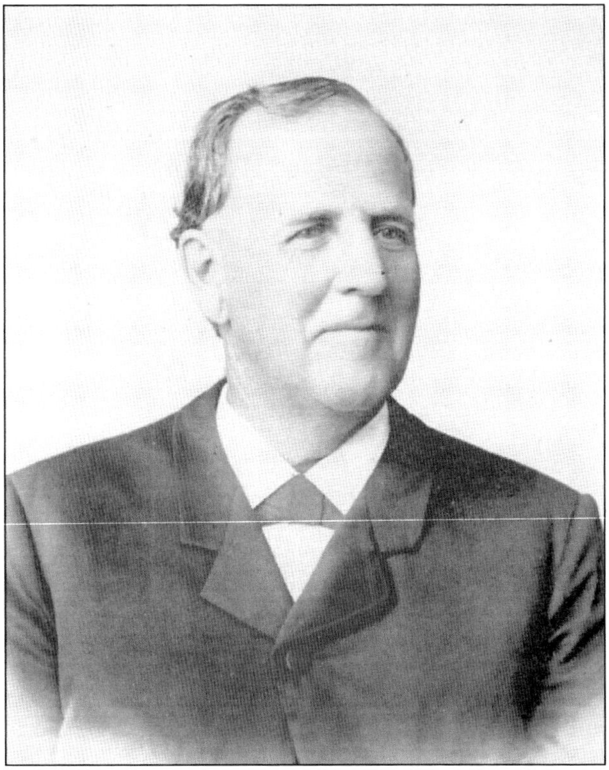

Asa Shuler (left) and John W. Benninghofen (right) formed a partnership to operate a small woolen mill on Fourth Street along the hydraulic in 1858. For several years prior to 1858, Shuler had been a partner with several other men in a similar undertaking at the same location. Benninghofen bought out Shuler's partners, and the company was known as the Miami Woolen Mills. In 1864, the mill began making felt sheeting needed by papermaking firms; these products became the primary product line for the company. The partnership passed down three generations of family members for more than 100 years and remained intact until 1967, when the company was sold to the Orr Felt Company of Piqua, Ohio.

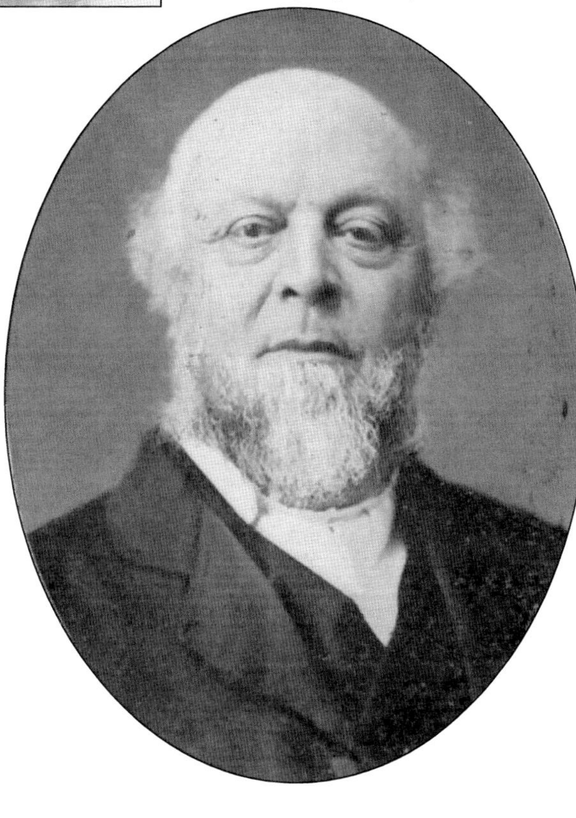

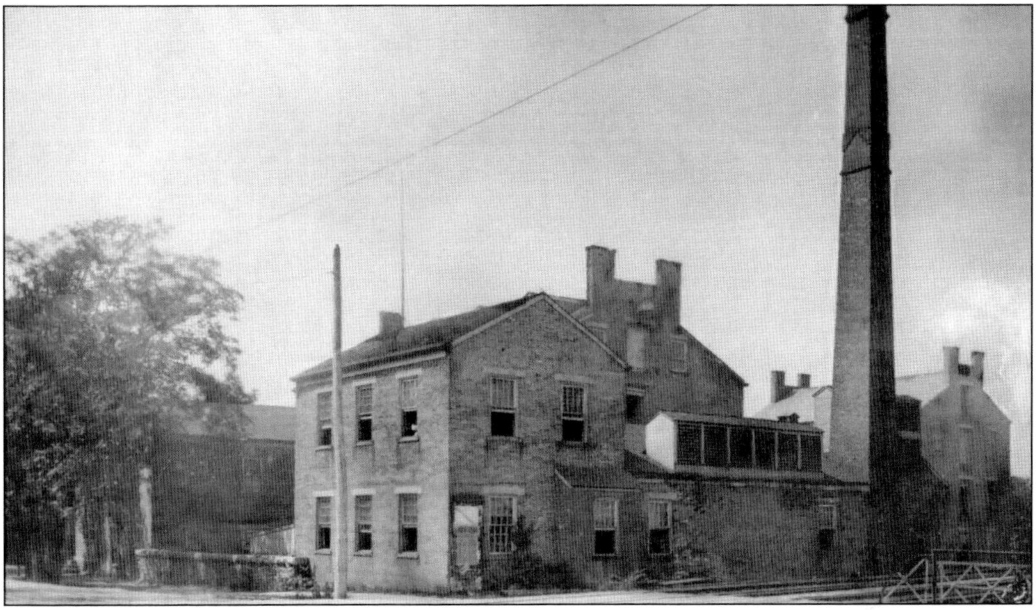

This 1893 photograph shows the second facility operated by Shuler and Benninghofen's Miami Woolen Mills. It was located on the northeast corner of Heaton and Lowell Streets, north of the Hooven, Owens, Rentschler Corliss Engine Works. A fire in 1865 had destroyed the first mill the firm built at this location in 1862. This photograph shows the replacement mill that was built in 1866.

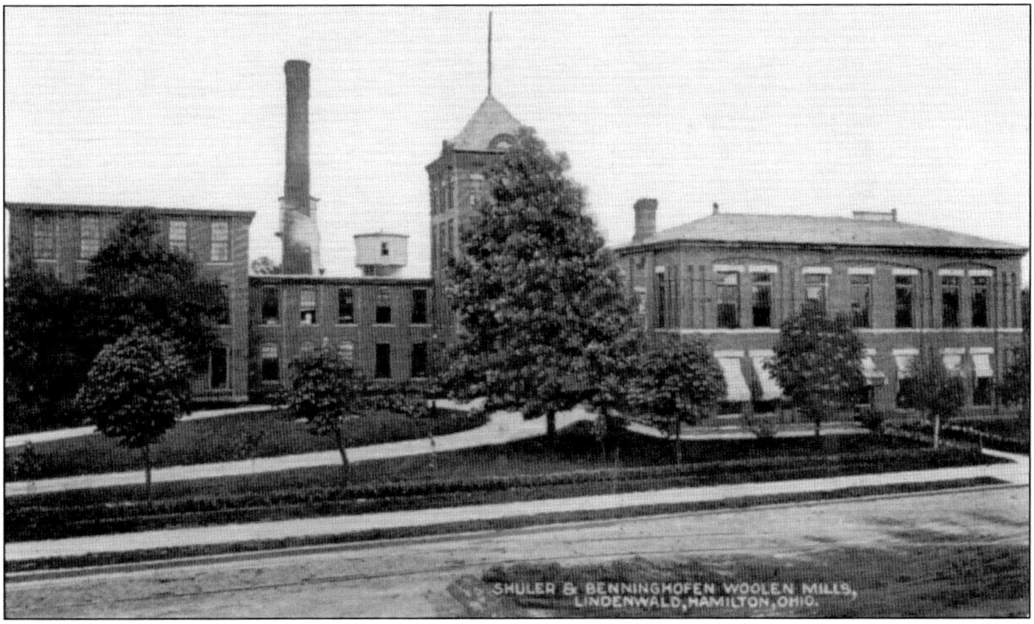

When this new plant opened on Pleasant Avenue in 1894, the Shuler and Benninghofen mill helped trigger the development of the Lindenwald area. The company's growth in business and the technological changes in the textile industry contributed to the partners' decisions to expand their factory in 1905, again in 1912, and once more in 1921.

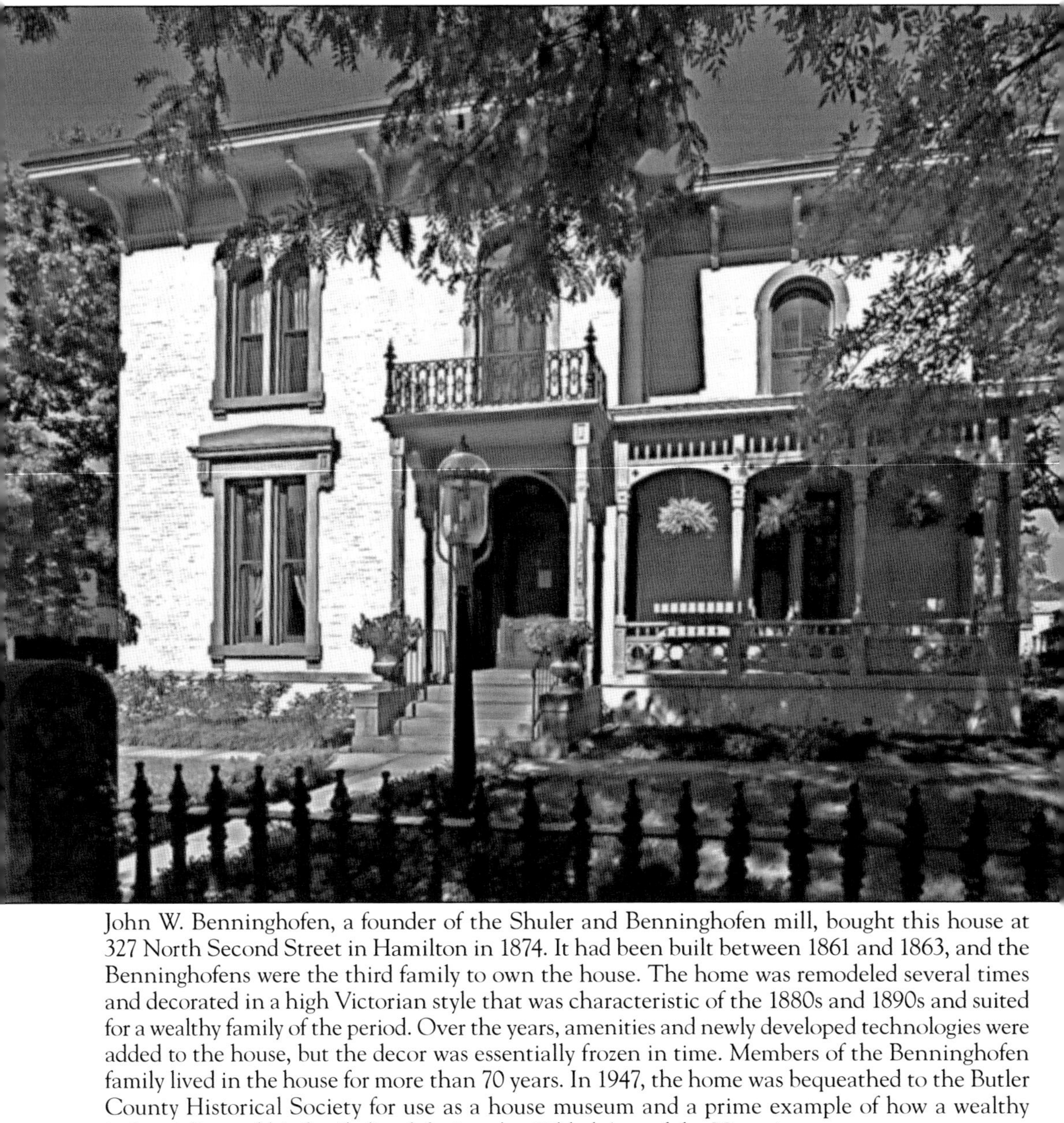

John W. Benninghofen, a founder of the Shuler and Benninghofen mill, bought this house at 327 North Second Street in Hamilton in 1874. It had been built between 1861 and 1863, and the Benninghofens were the third family to own the house. The home was remodeled several times and decorated in a high Victorian style that was characteristic of the 1880s and 1890s and suited for a wealthy family of the period. Over the years, amenities and newly developed technologies were added to the house, but the decor was essentially frozen in time. Members of the Benninghofen family lived in the house for more than 70 years. In 1947, the home was bequeathed to the Butler County Historical Society for use as a house museum and a prime example of how a wealthy industrialist and his family lived during the Gilded Age of the Victorian era.

Shuler and Benninghofen purchased this 762-inch loom from the De Laval Separator Company of Poughkeepsie, New York, in 1957. At the time, it was considered to be one of the largest looms in the world. The 17,000-pound machine was manufactured in Germany and installed on the second floor of the company's mill in Lindenwald.

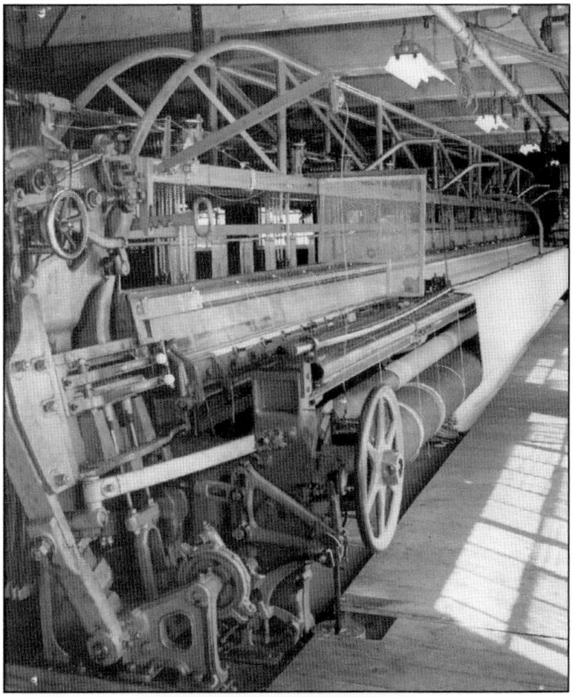

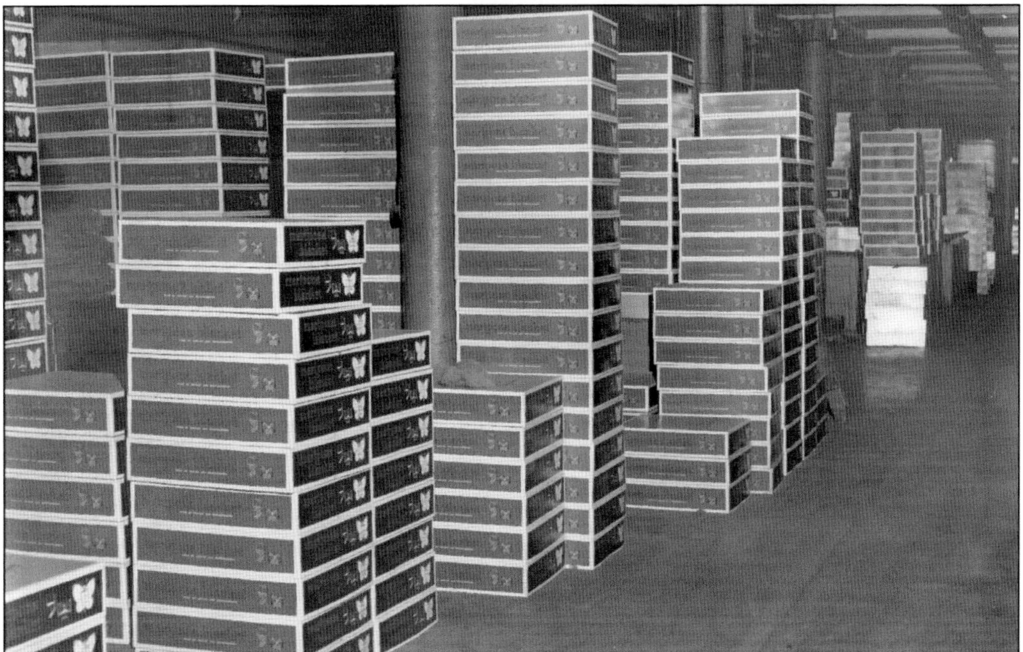

Hundreds of Shuler and Benninghofen's Mariposa-branded wool blankets are stacked and ready for shipment to distributors and merchants across the world in this c. 1958 photograph. As many as 125,000 blankets were made at the mill each year. Several other products were manufactured by the company, including Indian-pattern, single color, Hudson Bay pattern, and two-color striped blankets, wool clothing, and papermaker felts.

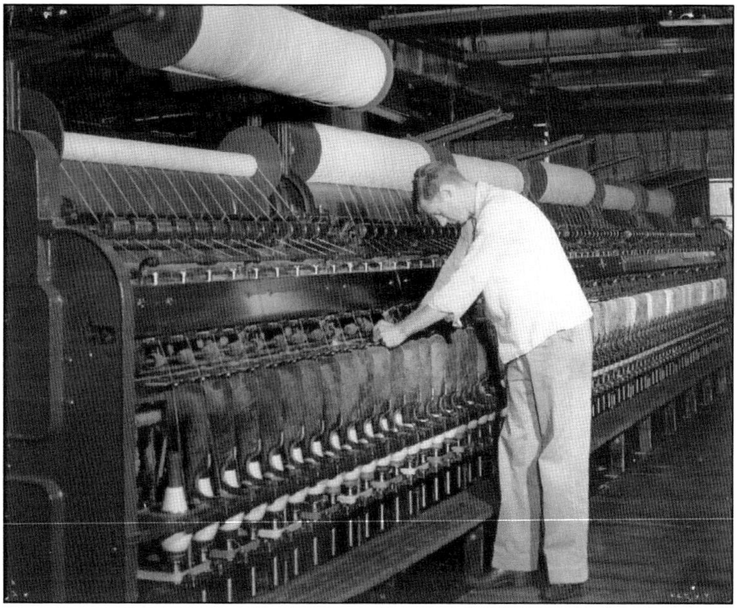

One of the Shuler and Benninghofen mill's spinning machines is seen on the second floor of the factory around 1960. Chris Johnson, the foreman of the spinning room, examines the operation of the machinery. The staff's attention to details and the company's manufacturing processes contributed to the mill making products of the highest quality.

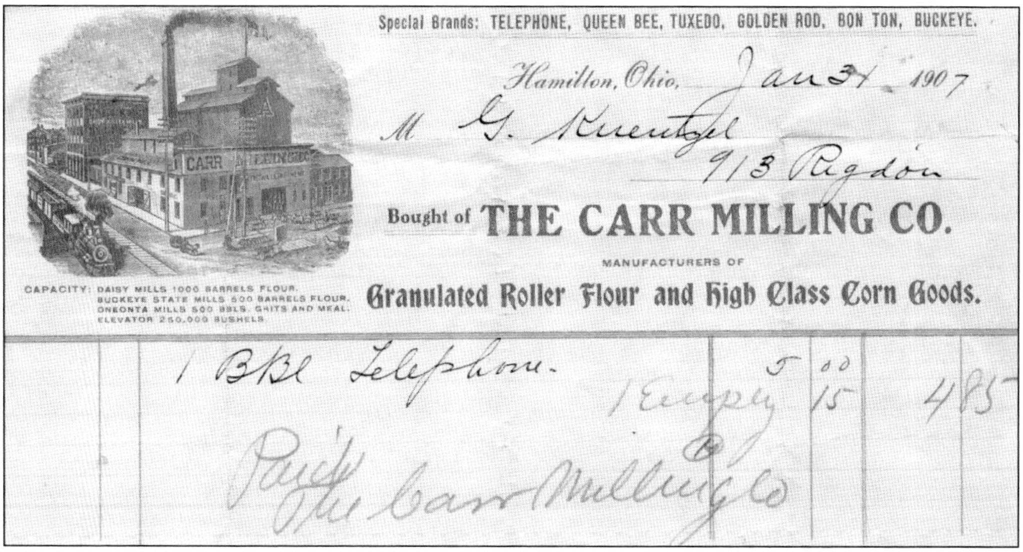

Special Brands: TELEPHONE, QUEEN BEE, TUXEDO, GOLDEN ROD, BON TON, BUCKEYE.

Hamilton, Ohio, Jan 31 1907

M G. Kuentzel

913 Rigdon

Bought of THE CARR MILLING CO.

MANUFACTURERS OF

Granulated Roller Flour and High Class Corn Goods.

CAPACITY: DAISY MILLS 1000 BARRELS FLOUR. BUCKEYE STATE MILLS 500 BARRELS FLOUR. ONEONTA MILLS 500 BBLS. GRITS AND MEAL. ELEVATOR 250,000 BUSHELS.

1 BBl Telephone. 5 00

1 Empty 15 4 85

Paid The Carr Milling Co

The Carr Milling Company grew out of the Carr and Brown flour mill that had started in 1868 and operated its mill and grain elevator at the corner of High and Fourth Streets. The company was incorporated in 1897. This 1907 receipt includes an image of the mill and shows that a 200-pound barrel of "Telephone" flour cost a customer $5. The firm was dissolved in 1939. It was the last of the old flour mills in Hamilton.

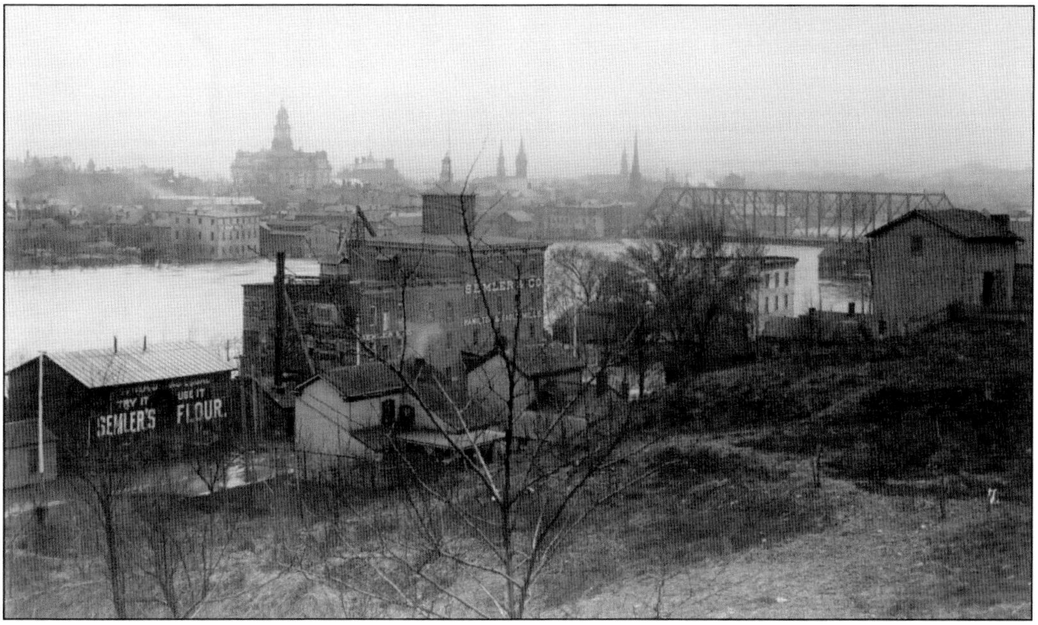

John Semler's mill was built in 1878 and rebuilt after major fires several times. Semler and Company was incorporated in 1897 and made the Mayflower and Point Lace brands of flour at the mill on North B Street. The mill was known as the Hamilton and Eagle Rolling Mills. This photograph of the mill was taken during the 1898 Great Miami River flood. The mill was sold in 1909; by 1912, it was known as the Prince Milling Company.

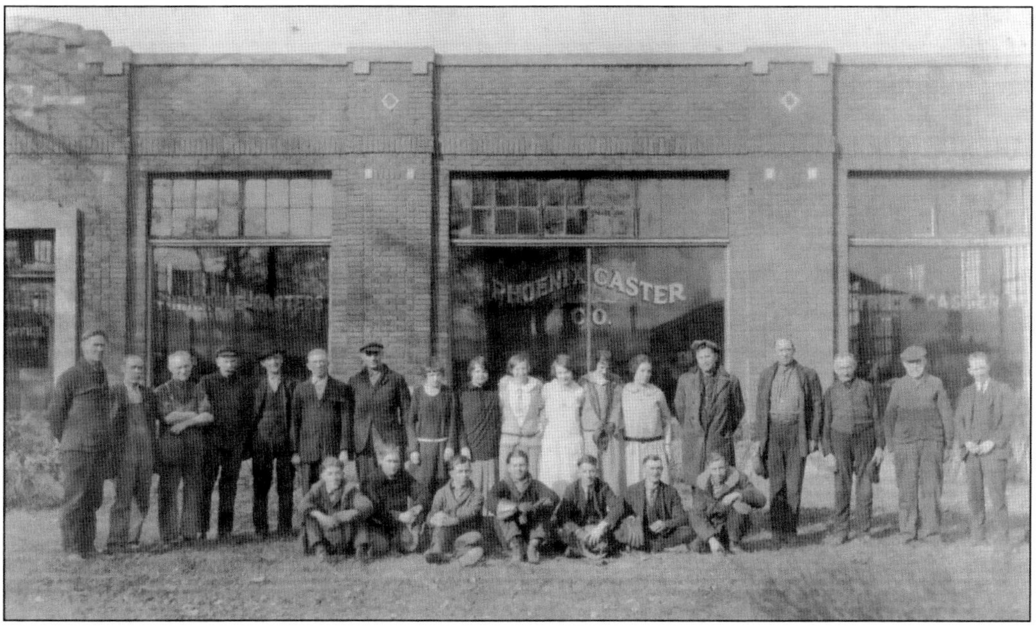

In 1874, Alexander Martin invented a caster that could move furniture without ruining carpeting. He won the support of Henry Sohn and George A. Rentschler, who began making the casters in 1877 at their foundry. In 1884, they incorporated the Phoenix Caster Company to produce an extensive line of furniture and truck casters. This undated photograph shows some of the firm's employees outside the company's building, which was the former Republic Motor Car plant.

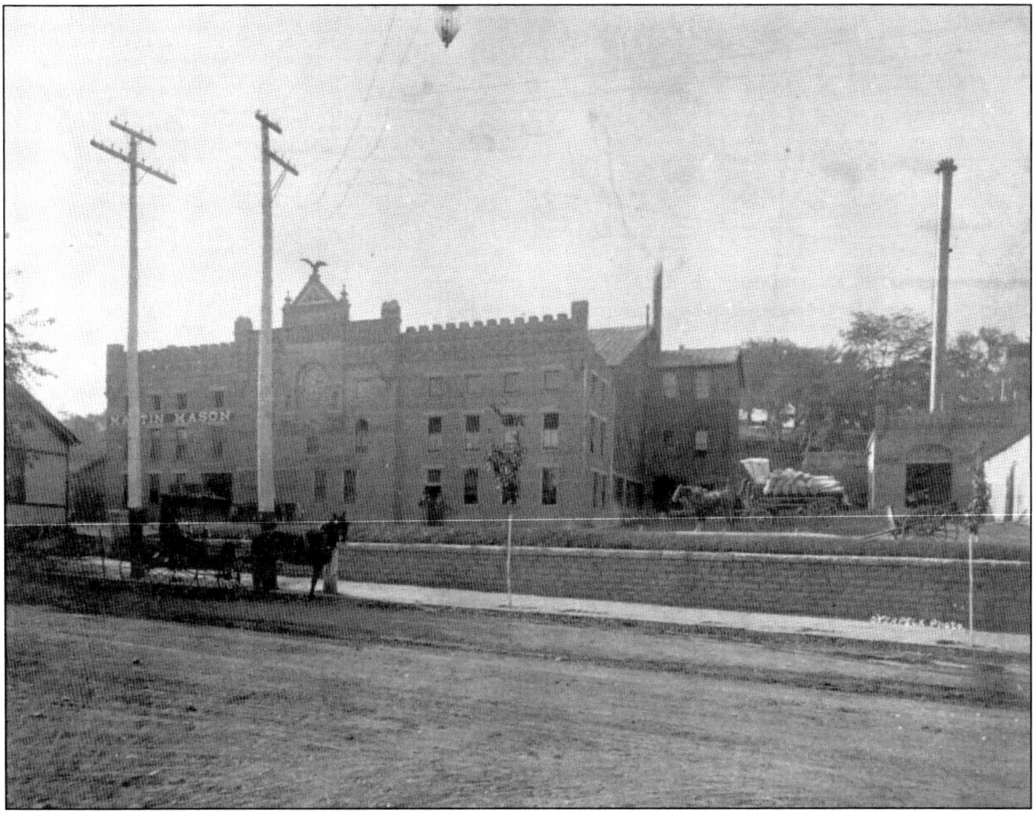

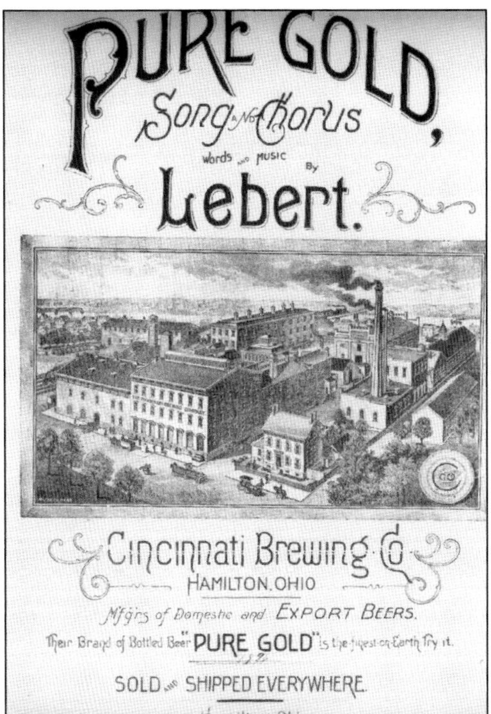

Several Hamilton companies brewed beer and distilled alcohol from the earliest days in the city. Beer brewers began using the building shown above to make their products in 1852. Martin Mason acquired sole ownership of the property in 1887. Known as the Eagle Brewery, the firm was actually incorporated as the Martin Mason Brewing Company in 1896. During Prohibition, the company remained in business by making near beer, but it went out of business in 1941. The brewery building was razed in 1977. Shown left is the sheet music for the "Pure Gold" theme song, written for the Cincinnati Brewing Company. Peter Schwab became sole owner of the firm in 1874 and incorporated it in 1882. Brewing ceased in 1919, and the company was dissolved in 1923. The facility was an ice plant until 1969. In 1977, the City of Hamilton opened a police headquarters and court on the land.

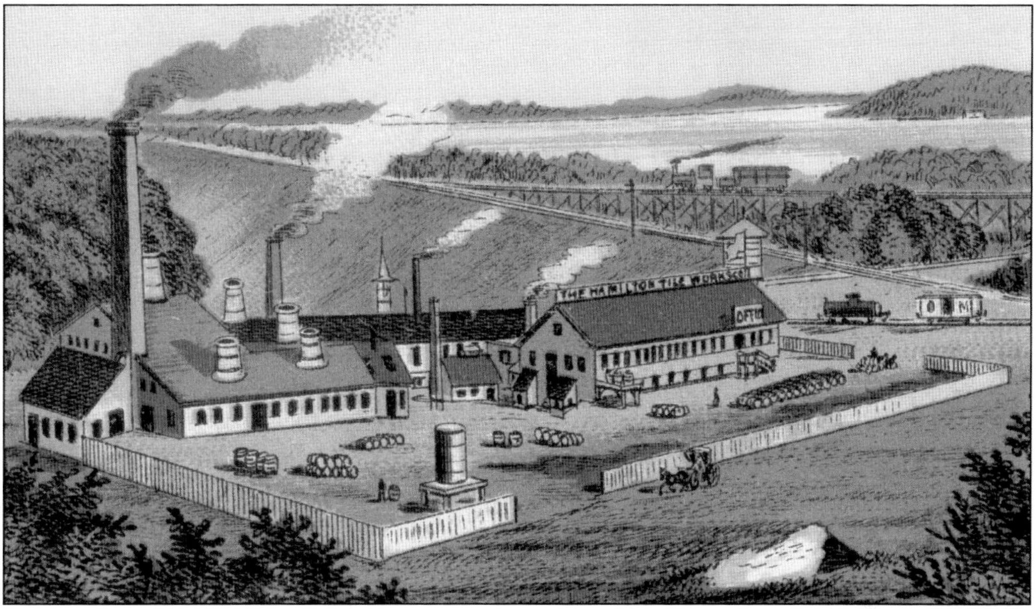

The roots of the Hamilton Tile Works go back to 1879, when George A. Rentschler and Joseph B. Hughes formed the Royal Pottery Works. Hughes withdrew from the company in 1881, and Rentschler leased the property to Adolph Metzner for use as a tile-making operation beginning in 1883. His business, incorporated as the Hamilton Tile Works in 1884, manufactured floor and wall tiles, decorative tiles for fireplaces, and mantels. The 1888 drawing above depicts his factory. Even though the company produced high-quality enameled tiles with beautiful designs and excellent glazing, it failed in 1897. A new ownership group including George A. Rentschler and Henry Sohn reincorporated the operation as the Ohio Tile Company in 1901. However, they too failed, and the business was dissolved in October 1911. Shown below is all that remained of the tile company after it was destroyed by the 1913 Great Miami flood.

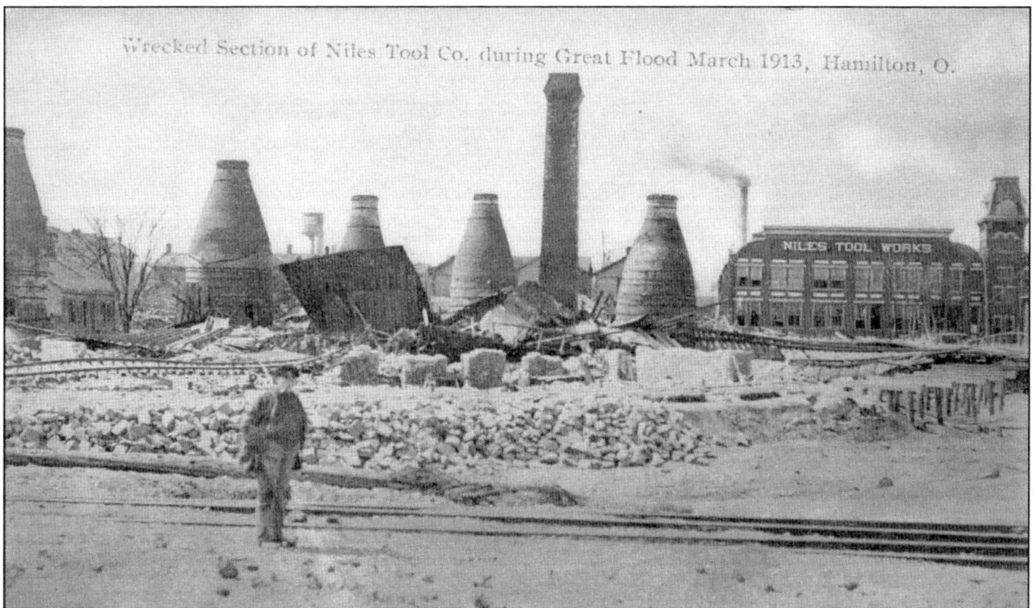

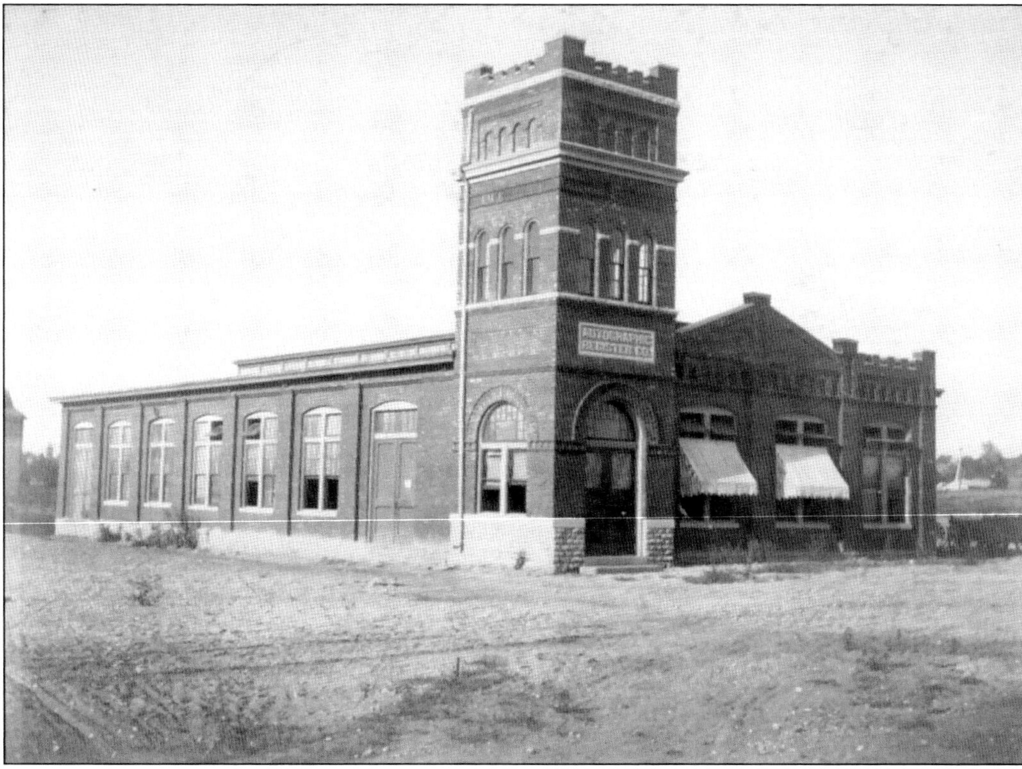

The Hamilton Autographic Register Company was formed in 1883 as a partnership between Albert Krauth and Peter and Christian Benninghofen. It operated out of a plant along the hydraulic on North Third Street. The company's factory on Symmes Avenue in Lindenwald (above) was built in 1894, seven years after the firm was incorporated. The primary product made by the company was the autographic register, a device that enabled a business to write multiple copies of an invoice or bill of sale at one time. An example of one of the models, the Full Record Autographic Register, is shown below. The device was ubiquitous throughout much of the 1900s. The company survived nearly a century until it was dissolved in 1980.

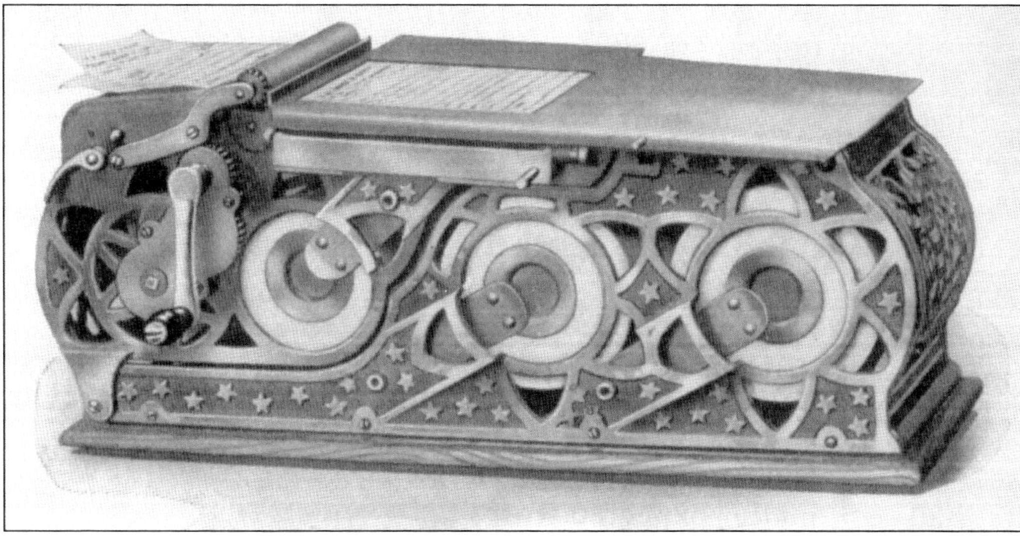

Krauth and Benninghofen, another partnership between Albert Krauth and Christian and Peter Benninghofen, began in 1883. The firm was an adjunct business tied to the Hamilton Autographic Register Company. Albert Krauth, seen right in a 1916 portrait, was the patentee of the autographic register and the machinery needed to manufacture it. He also owned patents for the Hamilton Music Stands, an example of which is shown left. Krauth and Benninghofen manufactured all sorts of music stands, guitar capos, and other music-related items. The company was incorporated in 1964, when the firm built a plant on Symmes Road. The Benninghofen heirs sold their interest in the firm in 1967, and it merged with a Chicago music-instrument company and then became part of the James David furniture manufacturer in 1976. The rights to make the music stand product line were sold in 1982, and production was continued by the Hamilton Stands Company.

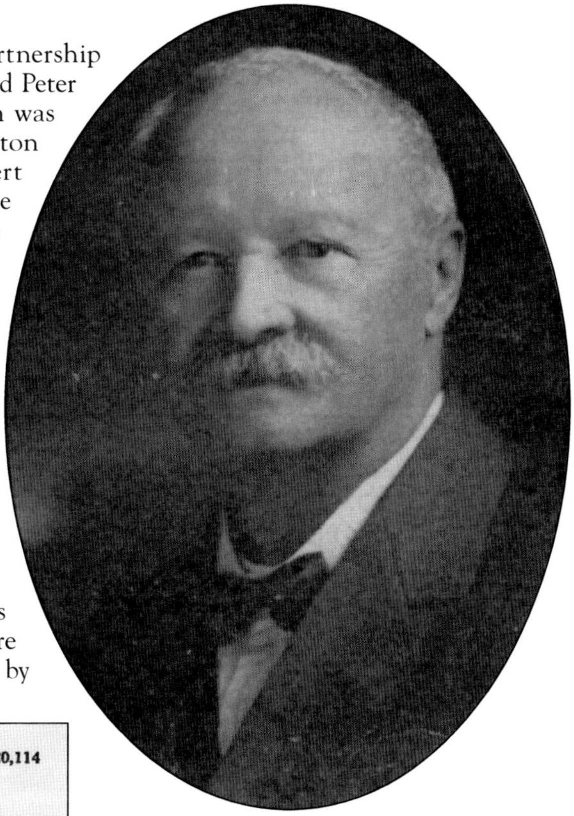

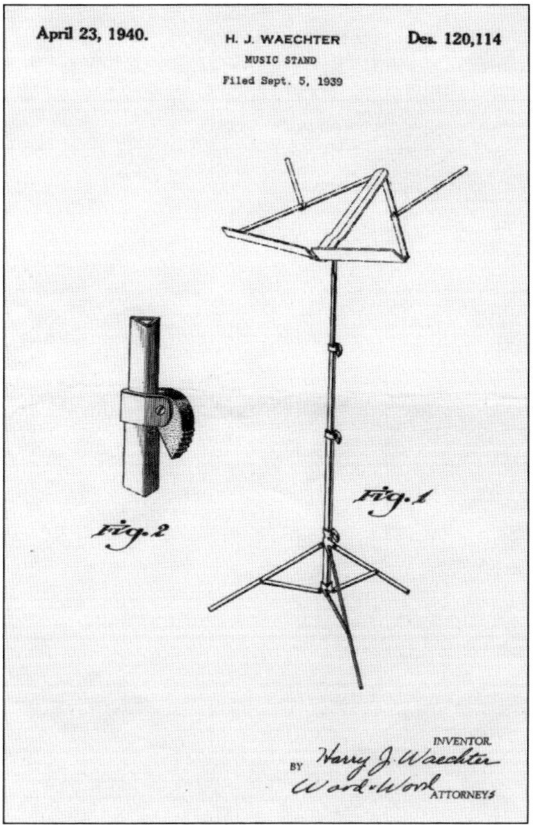

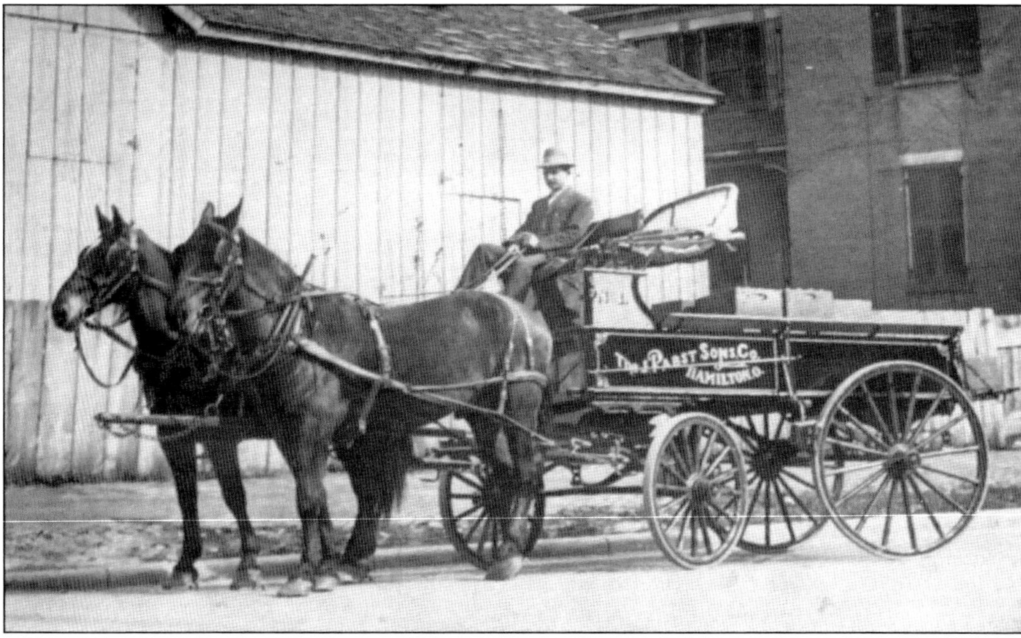

Edward L. Keller, salesman for the J. Pabst and Sons Company, appears ready to make deliveries to his customers in the above photograph from 1908. Jacob Pabst Sr. organized his beverage company in 1888 and incorporated it in 1899. The firm manufactured and bottled a variety of soft drinks, including fruit-flavored sodas, sarsaparilla, root beer, Vichy-still water, champagne mist, and a very popular ginger ale. The company also made cider, vinegar, and distilled-water products. Pabst's brewing and bottling works were in the building seen below in a photograph taken in 1928. It was located on Central Avenue near Pershing Avenue. The company became the first local bottler of Pepsi-Cola in 1910. The Canada Dry Bottling Company of Cincinnati purchased Pabst and Sons in 1965. The business was closed in 1973 and completely dissolved in 1981.

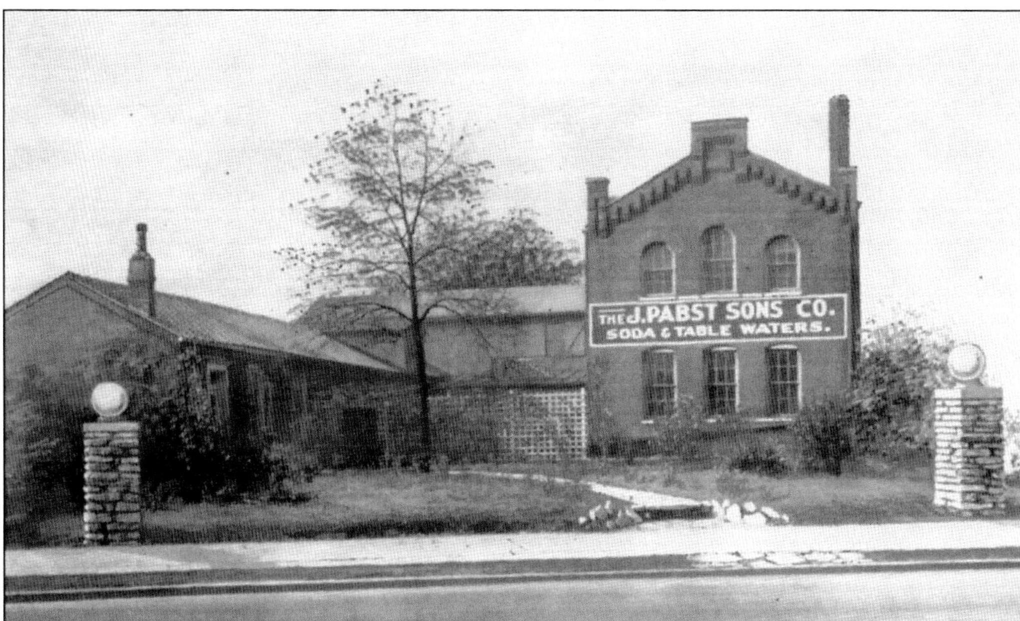

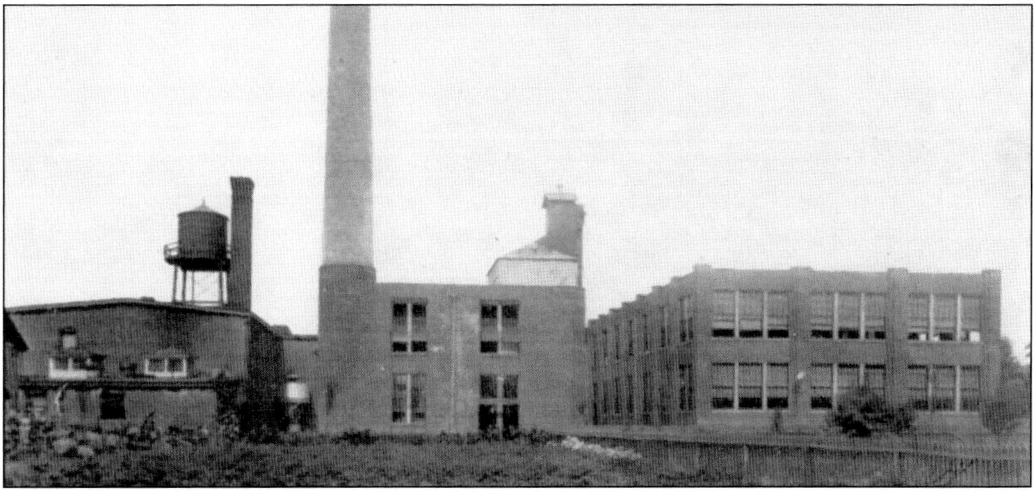

The Miami Valley Knitting Mills began as a small garment-making operation in 1890. The firm produced ladies, children's, and infant's underwear, all usually manufactured out of ribbed fabric. The company was founded by Henry Niederauer, incorporated in 1911, and remained a family operation throughout its history. This photograph shows the company's factory, located on the northeast corner of South Ninth and Hanover Streets.

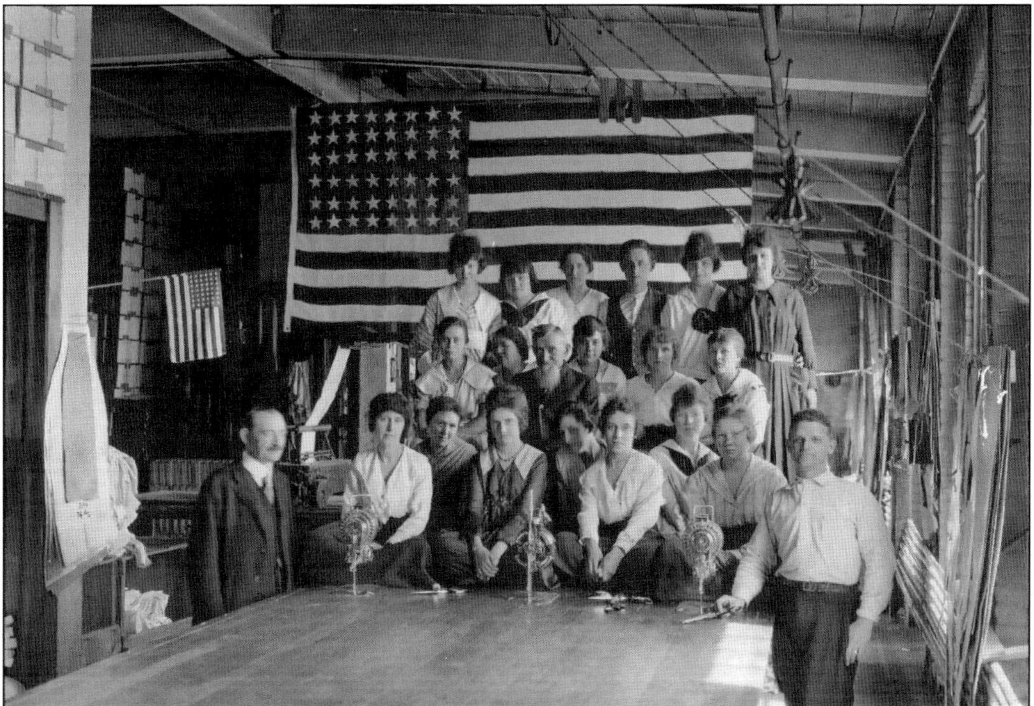

This April 1918 photograph shows a group of unidentified Miami Valley Knitting Mills employees in one of the company's cutting rooms. Workers would lay out the knitted fabric and use the forms stored along the walls to cut the material to the size needed for a garment. One of the company's most successful products was the union suit, the one-piece, form-fitting, long-sleeved underwear that buttoned up the front from the groin to the neck.

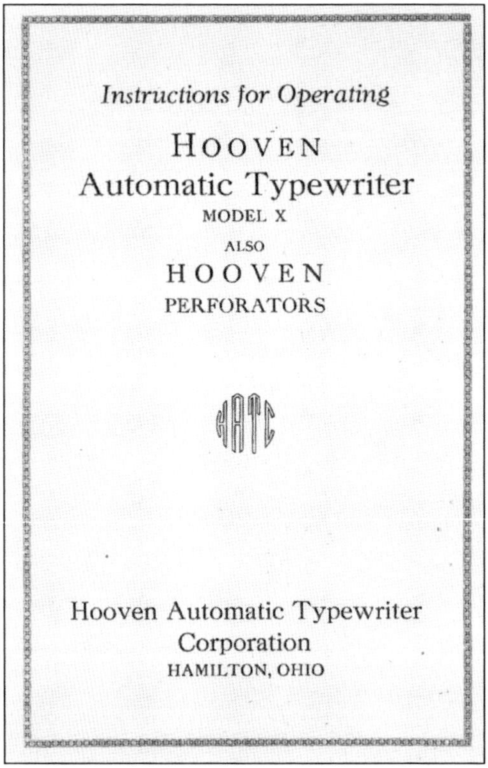

Instructions for Operating

HOOVEN
Automatic Typewriter
MODEL X

ALSO

HOOVEN
PERFORATORS

Hooven Automatic Typewriter
Corporation
HAMILTON, OHIO

This instruction manual explained how users could produce hundreds of identical typed letters, each one personalized to a particular recipient. The device was introduced in Cincinnati in 1912 by the National Automatic Typewriter Company, headed by C. Earle Hooven, grandson of the founder of Hooven, Owens, Rentschler. In 1916, production shifted to Hamilton, and by 1919, the company's name had been changed to Hooven Automatic Typewriter Company.

The Hooven Automatic Typewriter required two machines for its operation. The perforator had a normal four-row typewriter keyboard used to type a letter and, in the process, punch a series of holes into a paper roll similar to a perforated roll on a player piano. The punched paper roll was put into a device that controlled a typewriter and allowed an operator to insert personal details in a document that was otherwise exactly the same form letter.

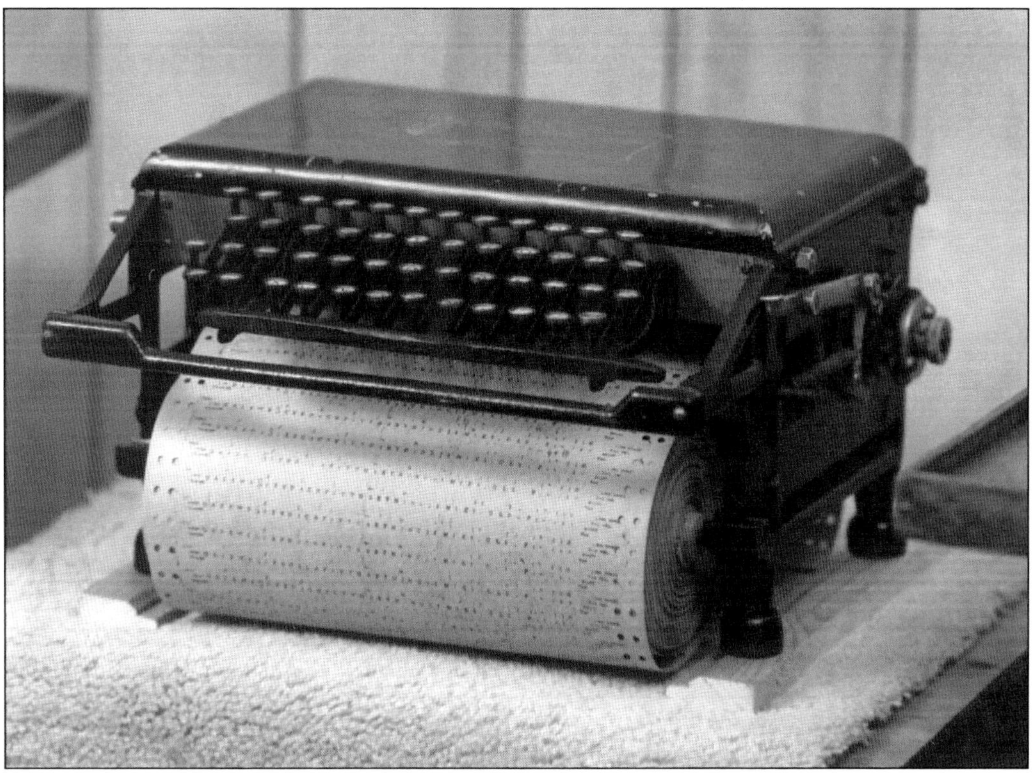

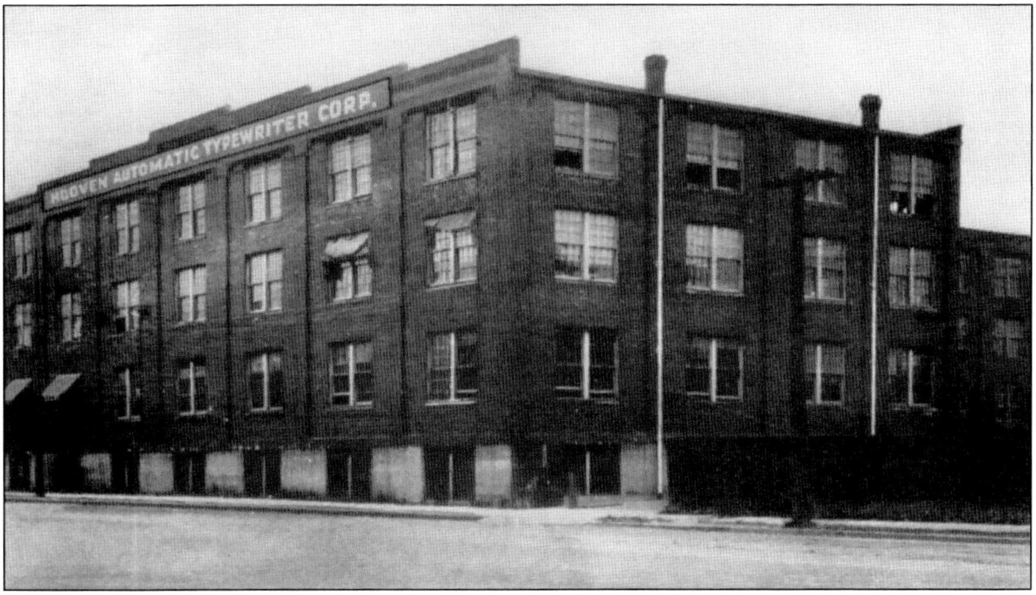

In 1919, the Hooven Automatic Typewriter Company began manufacturing its perforator and other devices in this building, the former American Can Company factory on Central Avenue. Even though the Hooven was relatively expensive, it was popular until World War II. After the war, other companies developed electric typewriters, and the product became obsolete. Production of the automatic typewriter ceased in Hamilton when the company moved to Dayton, Ohio, in 1935 and became the Hooven-Dayton Corporation.

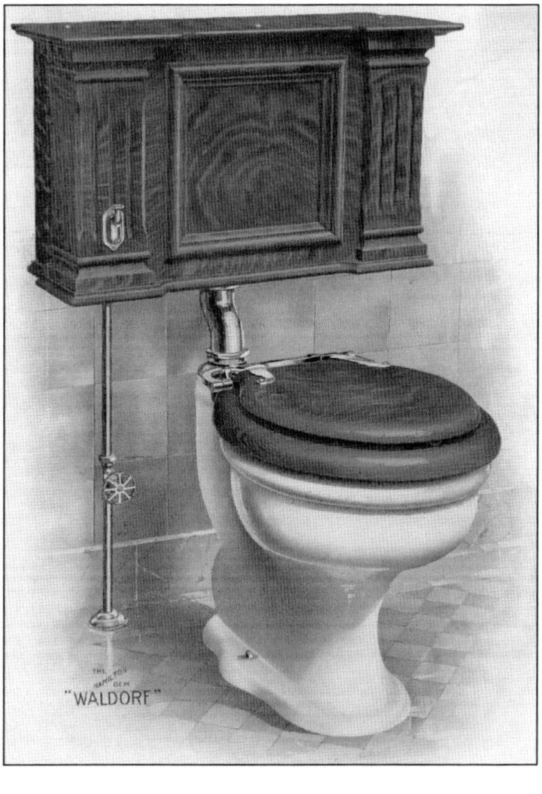

The Sanitary Manufacturing Company's Waldorf water closet, and as many as 99 other styles of "Hamilton Gems," were made at the firm's plant at Front and Dayton Streets. The firm was formed in 1900 to manufacture its trademarked toilet bowls, plumber's woodwork tanks, and seats. The company was sold to the Huntington Sanitary Manufacturing Company of West Virginia in April 1916. Within three months, the plant and equipment moved out of Hamilton.

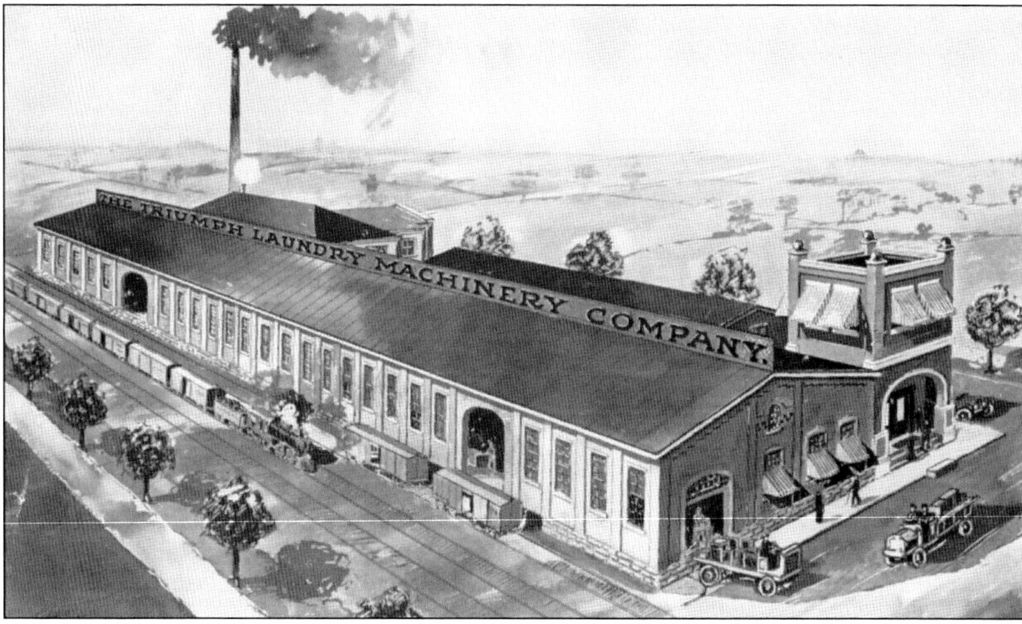

Archibald Davidson opened a boiler-manufacturing business around 1880 and began making traction engines as the Empire Machine Company in 1893. The office and works were located on Walnut Street, west of South Seventh Street. The building, seen above in an artist's rendering, was only 60 feet by 120 feet in size. By 1910, the company was incorporated as the Triumph Laundry Machinery Company. The firm manufactured a full line of laundry equipment used by steam laundries, hotels, and institutions. The products the company made included washers, extractors, several types of ironers, presses, and the Triumph flat work ironer (below), patented in 1910. The company ceased production in 1919, and its property on Walnut Street was purchased by the Graf Motor Car Company.

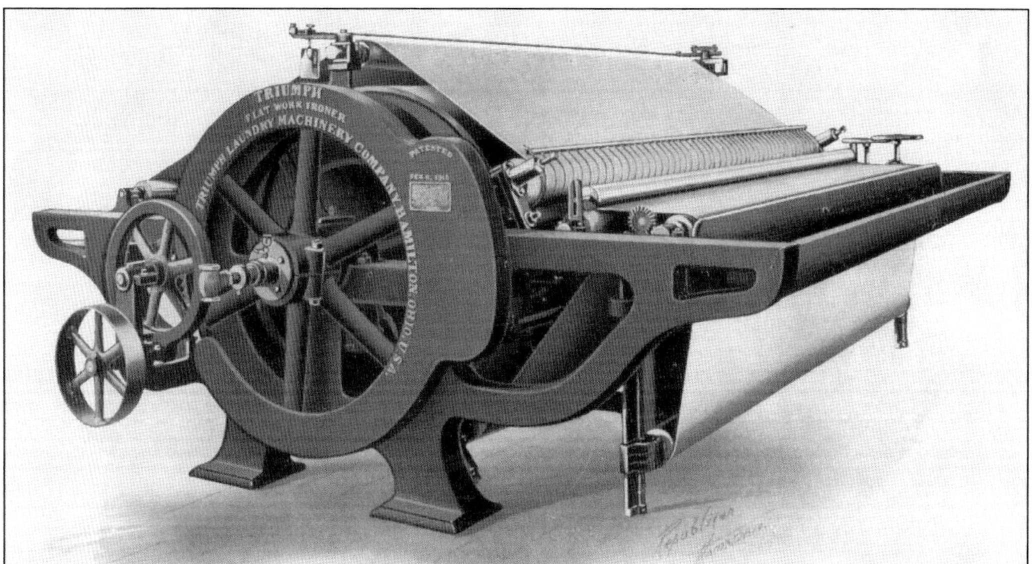

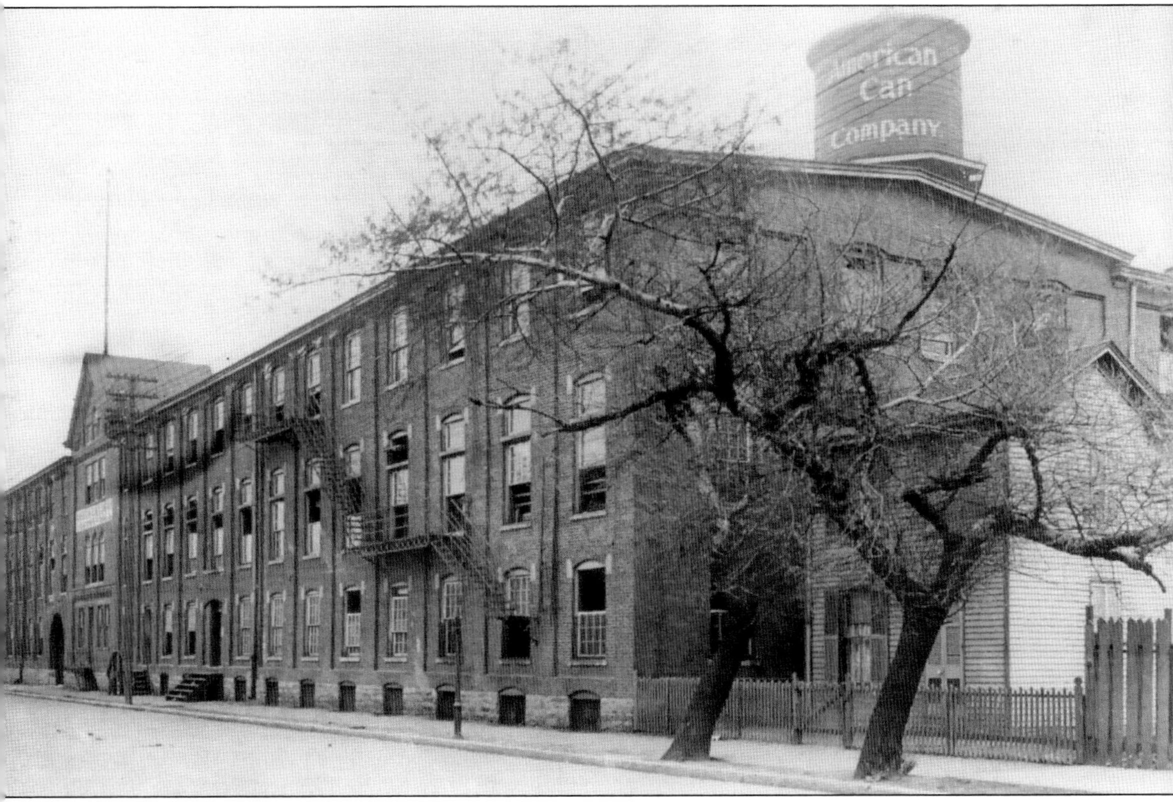

Albert Fisher moved his canning company from Cincinnati to Hamilton in 1885. The firm built this large three-story factory on North Third Street near Black Street, where it made tin cans and canned a variety of vegetables. By 1899, the company stopped its canning operations and focused exclusively on making tin cans, pails, and other forms of packaging. Over the years, the company manufactured a wide assortment of packaging for products such as candy, baking powder, talcum, tooth powder, lard, syrup, coffee, crackers, and spices. In 1901, the American Can Company was formed as a merger of 60 firms representing 123 factories across 11 states. The Albert Fisher Manufacturing Company was one of the 60 companies that merged into what was essentially a "tin can trust." The federal government sued to break up the trust in 1913. The Albert Fisher–American Can operation remained in business in Hamilton until 1931, when the company was acquired by the Metal Package Corporation of Delaware.

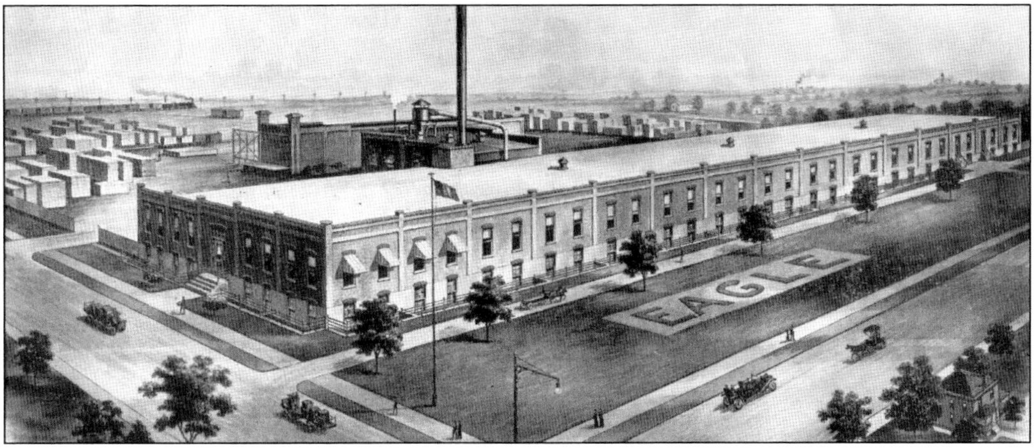

Nicholas Fritsch began the cooperage works that evolved into the Eagle Woodenware Company in Cincinnati in 1871. The firm was making barrels and kegs when it relocated to Zanesville, Ohio, in 1899. By the time the company relocated to Hamilton in 1912, it was making buckets, baskets, pails, tubs, mops, washing machines, and a variety of cleaning tools. In later years, the firm would create products such as galvanized steel pails with mop-wringing equipment. The company's new two-story factory, seen above in an artist's rendering, was located on Dayton Avenue east of North Erie Highway. The company prepared an exhibit (below) for the 1916 celebration of the 125th anniversary of the founding of Fort Hamilton. The display was in the industrial exposition's machinery hall at the Butler County Fairgrounds.

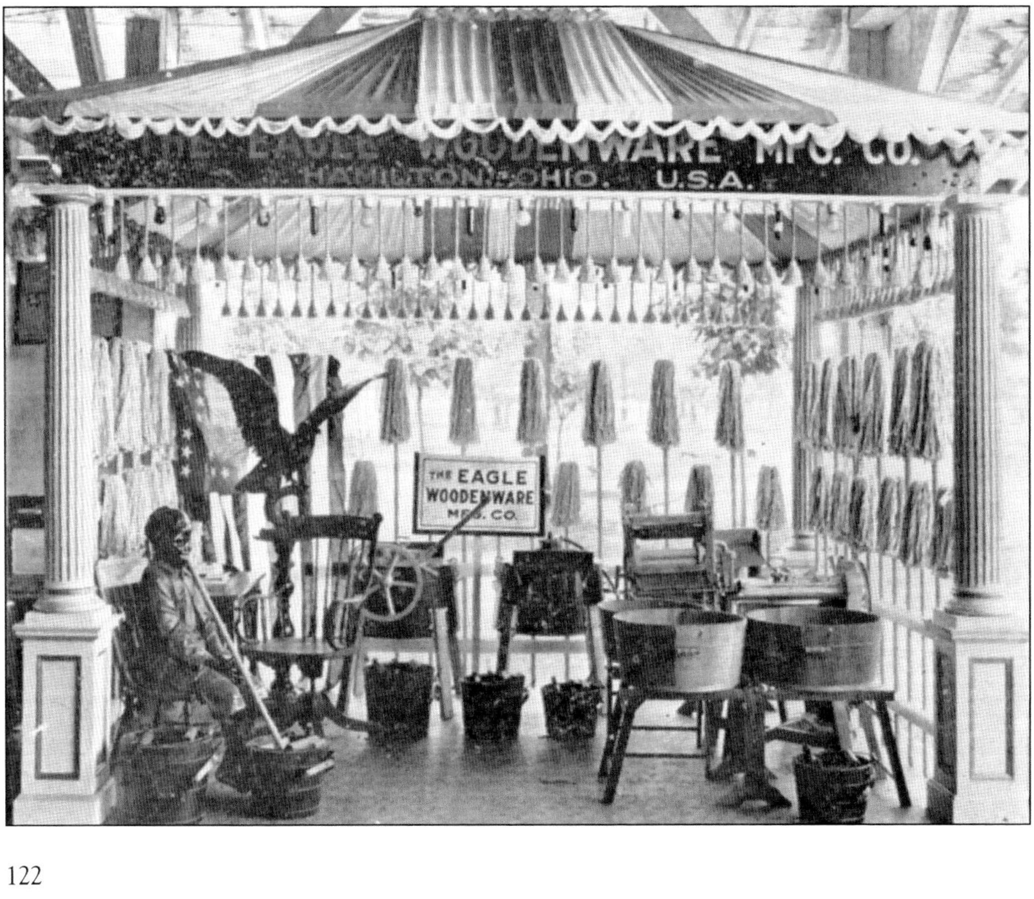

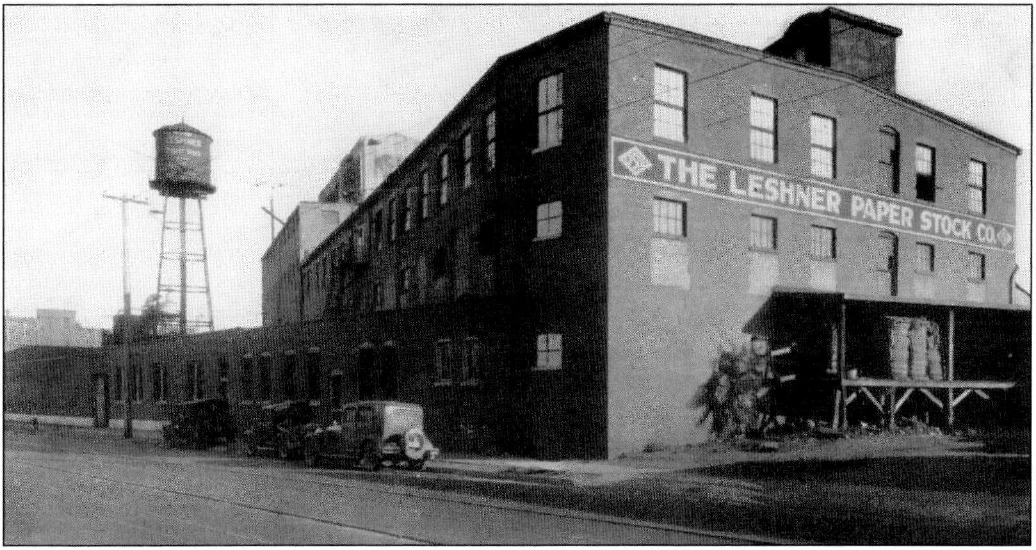

Nathan Leshner established the Leshner Paper Stock Company in 1916 to provide paper and rags, scrap materials, and roofing waste to paper manufacturers. The company was housed in a factory (above) that was part of a complex of buildings on Central Avenue. By 1925, Leshner had formed a partnership with the Philip Carey Company, which took full ownership of the firm upon his death in 1937. In 1947, a group of Leshner employees purchased the company from Philip Carey and renamed it the Leshner Corporation. The company continued acting as an agent for Philip Carey and sold towels, washcloths, cheesecloth, and many other textile products as a broker for mills such as Cannon and Dundee. These products were sold in the Leshner Cotton Mill store seen below.

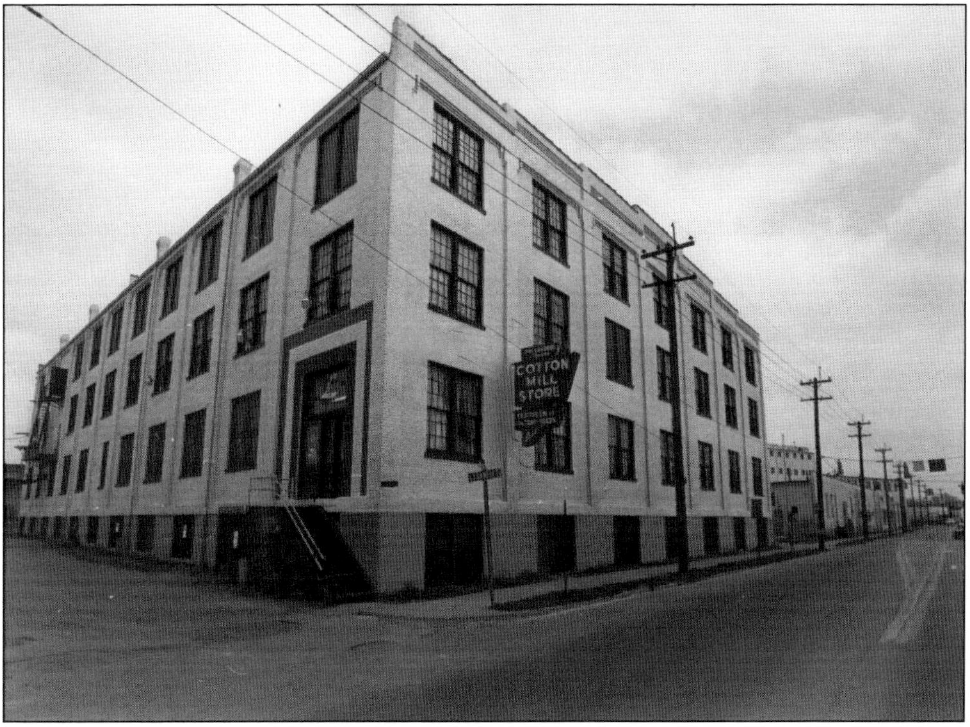

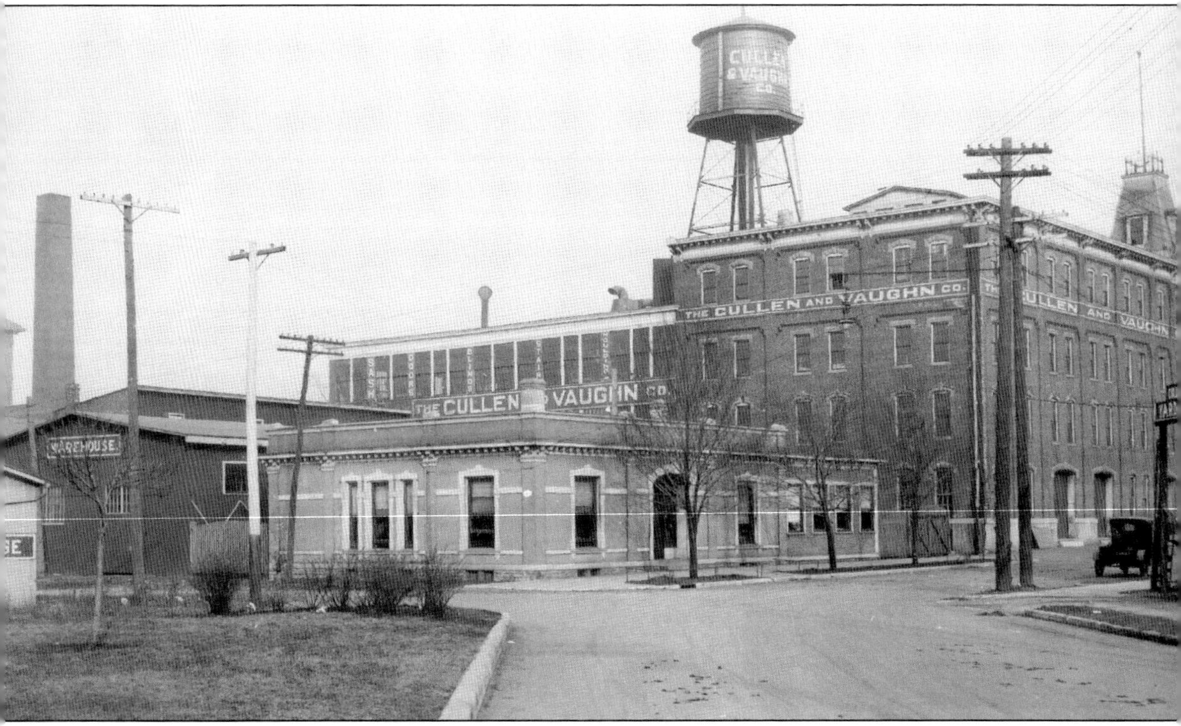

J. Webster Cullen and Frank K. Vaughn formed a successful construction business in 1911. The company acquired the Arent lumberyard in 1911 and the Semler & Company mill on North B Street in 1912. In October 1913, they acquired the Bender Company, the dominant construction company in the area. The Bender Company began as F.W. Bender and Sons in the early 1860s and then became the J.F. Bender & Brothers Company in 1867 before its name was changed again in 1908. Cullen and Vaughn erected many buildings, including the Hamilton YMCA, Van Buren School, Niles Tool Works additions, and several paper mills and foundry additions. The firm's four-story factory, where the company manufactured doors, molding, millwork, and other materials, is seen here in 1919. The facility, located at Fourth and Dayton Streets, was sold to the Beckett Paper Company in 1919. Cullen, son of Niles Tool Works president James K. Cullen, left the company in 1916 and died from injuries he received during World War I military service in 1919. Frank Vaughn formed the Vaughn Building Company in 1916. His firm built Hamilton's Ford Tractor plant in 1919 and was successful for many years until it was dissolved in 1985.

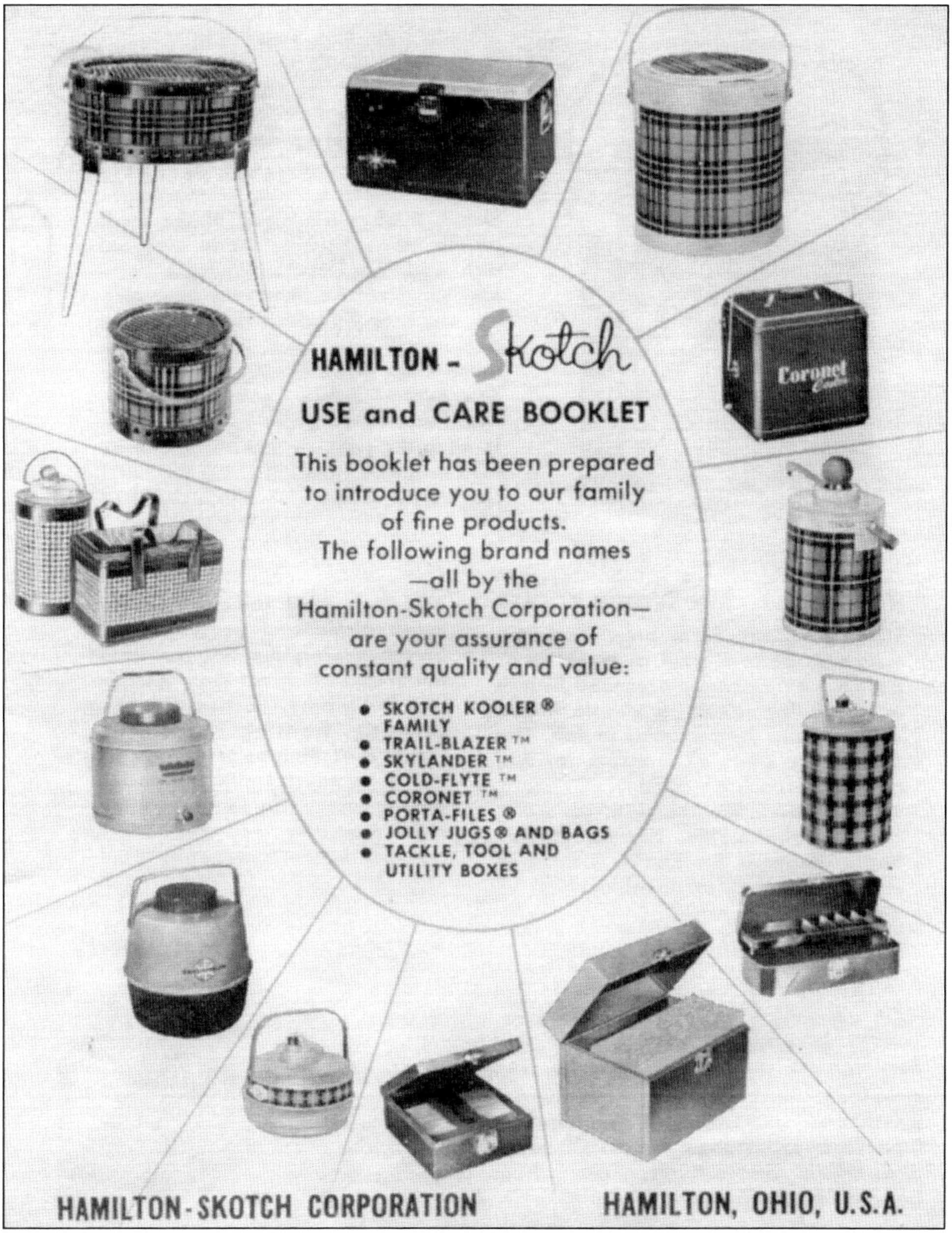

The Hamilton Sheet Metal Company was founded in 1910 by Louis Piker and Meyer Fogel to manufacture mailboxes. The firm moved to a plant in Lindenwald in 1925 and, by the 1940s, was making toolboxes, medicine cabinets, and other metal items. The company's name was changed to Hamilton Metal Products Company in 1951. During the 1950s, New York designer Petra Cabot created a variety of products that featured a red, black, and yellow tartan plaid look. This line of coolers, some of which are seen on the cover of this booklet, was quite popular with consumers. The firm moved to a plant on South Erie Boulevard in 1957 and promoted itself as the Hamilton Skotch Corporation to capitalize on its popular line of products. By 1970, however, the company was no longer located in Hamilton, and some of its plaid-design products were being made in Ottawa, Kansas.

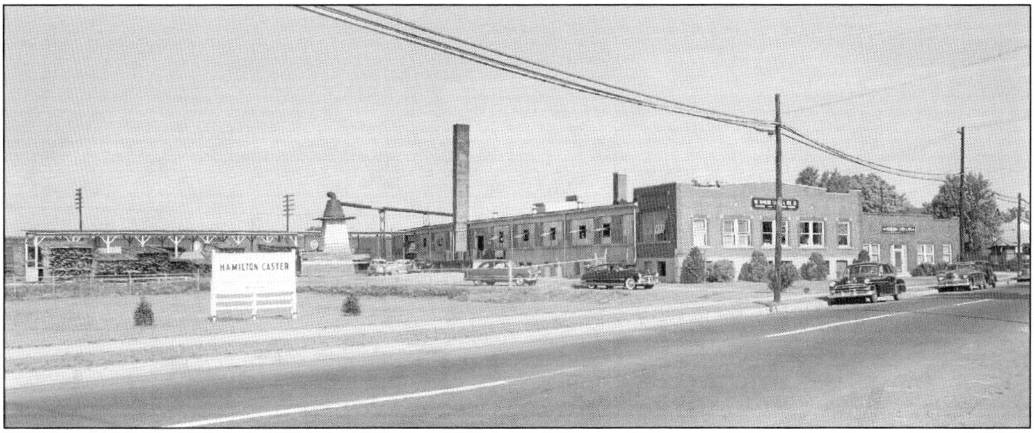

In 1907, John A. Weigel bought the patents for a nonclogging, double-wheeled industrial caster and began a small operation to manufacture them. By 1920, Weigel's company, incorporated as the Hamilton Caster & Manufacturing Company in 1910, built a factory on Dixie Highway. This view of the enlarged plant dates from the 1940s. The company made other additions and improvements to its factory as the diversity of the firm's product lines increased. (Courtesy of Hamilton Caster & Manufacturing Co.)

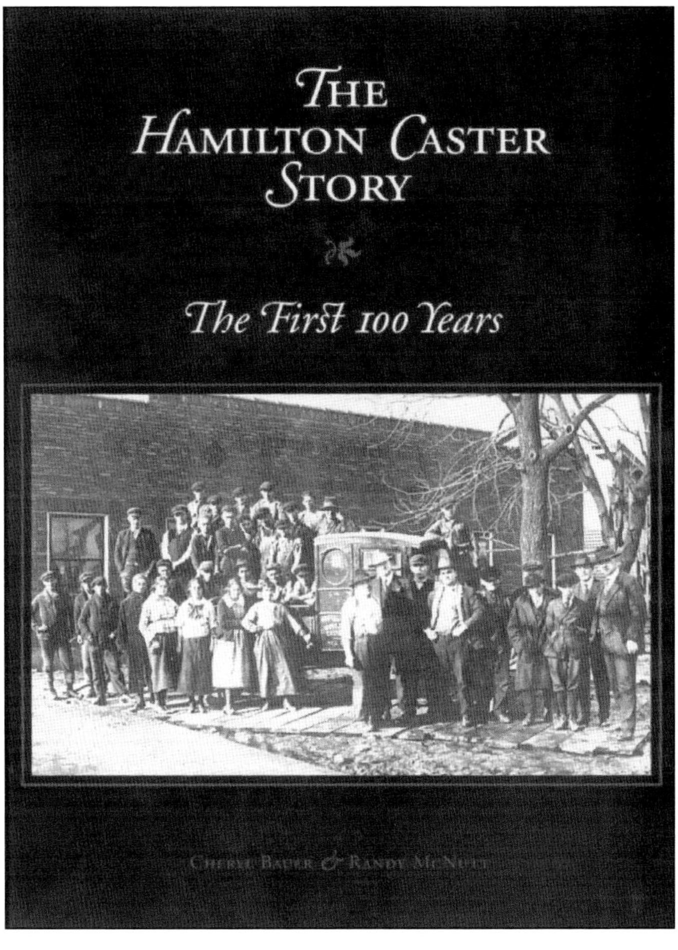

For more than 100 years, Hamilton Caster & Manufacturing Company has remained a family-owned business. It is today the oldest manufacturing firm still operating in Hamilton. In 2007, the company published a history of the firm, written by Cheryl Bauer and Randy McNutt. The photograph on the book cover dates from 1920 and shows the firm's employees in front of the company's new plant on Dixie Highway. (Courtesy of Hamilton Caster & Manufacturing Co.)

INDEX

Consistent with our mission to preserve history on a local level, this book was printed in South Carolina on American-made paper and manufactured entirely in the United States. Products carrying the accredited Forest Stewardship Council (FSC) label are printed on 100 percent FSC-certified paper.

MADE IN THE USA